Practical Machinery Safety

Other titles in the series

Practical Data Acquisition for Instrumentation and Control Systems (John Park, Steve Mackay)

Practical Data Communications for Instrumentation and Control (Steve Mackay, Edwin Wright, John Park)

Practical Digital Signal Processing for Engineers and Technicians (Edmund Lai)

Practical Electrical Network Automation and Communication Systems (Cobus Strauss)

Practical Embedded Controllers (John Park)

Practical Fiber Optics (David Bailey, Edwin Wright)

Practical Industrial Data Networks: Design, Installation and Troubleshooting (Steve Mackay, Edwin Wright, John Park, Deon Reynders)

Practical Industrial Safety, Risk Assessment and Shutdown Systems for Instrumentation and Control (Dave Macdonald)

Practical Modern SCADA Protocols: DNP3, 60870.5 and Related Systems (Gordon Clarke, Deon Reynders)

Practical Radio Engineering and Telemetry for Industry (David Bailey)

Practical SCADA for Industry (David Bailey, Edwin Wright)

Practical TCP/IP and Ethernet Networking (Deon Reynders, Edwin Wright)

Practical Variable Speed Drives and Power Electronics (Malcolm Barnes)

Practical Centrifugal Pumps (Paresh Girdhar and Octo Moniz)

Practical Electrical Equipment and Installations in Hazardous Areas (Geoffrey Bottrill and G. Vijayaraghavan)

Practical E-Manufacturing and Supply Chain Management (Gerhard Greef and Ranjan Ghoshal)

Practical Grounding, Bonding, Shielding and Surge Protection (G. Vijayaraghavan, Mark Brown and Malcolm Barnes)

Practical Hazops, Trips and Alarms (David Macdonald)

Practical Industrial Data Communications: Best Practice Techniques (Deon Reynders, Steve Mackay and Edwin Wright)

Practical Machinery Vibration Analysis and Predictive Maintenance (Cornelius Scheffer and Paresh Girdhar)

Practical Power Distribution for Industry (Jan de Kock and Cobus Strauss)

Practical Process Control for Engineers and Technicians (Wolfgang Altmann)

Practical Telecommunications and Wireless Communications (Edwin Wright and Deon Reynders)

Practical Troubleshooting Electrical Equipment (Mark Brown, Jawahar Rawtani and Dinesh Patil)

Practical Machinery Safety

David M. Macdonald BSc (Hons) Inst. Eng., Senior Engineer,
IDC Technologies, Cape Town, South Africa

Series editor: Steve Mackay

AMSTERDAM • BOSTON • HEIDELBERG • LONDON
NEW YORK • OXFORD • PARIS • SAN DIEGO
SAN FRANCISCO • SINGAPORE • SYDNEY • TOKYO

Newnes is an imprint of Elsevier

Newnes

Newnes
An imprint of Elsevier
Linacre House, Jordan Hill, Oxford OX2 8DP
200 Wheeler Road, Burlington, MA 01803

First published 2004

British Library Cataloguing in Publication Data
Macdonald, D.M.
    Practical machinery safety. – (Practical professional)
    1. Machinery – Safety measures   2. Machinery – Safety
    appliances   3. Industrial safety
    I. Title
    621.8'0289

Library of Congress Cataloguing in Publication Data
A catalogue record for this book is available from the Library of Congress

ISBN   0 7506 6270 0

For information on all Newnes publications
visit our website at www.newnespress.com

Typeset and edited by Integra Software Services Pvt. Ltd, Pondicherry, India
www.integra-india.com
Printed and bound in The Netherlands

TJ
1177
.M33
2004 – Apr05

Contents

Preface viii

1   Introduction to the machinery safety workshop 1
     1.1   Scope and objectives 2
     1.2   Machinery and controls 2
     1.3   Distinction between machinery and process safety control systems..... ...7
     1.4   International standards and practices...................................... ...8
     1.5   Introduction to hazards and risks...10
     1.6   Risk reduction...11
     1.7   The Alarp principle for tolerable risk 12
     1.8   Development example for a machinery safety system........................14
     1.9   The engineering tasks 19
     1.10   Benefits of the systematic approach.......................................22
     1.11   Conclusions 23

2   Guide to regulations and standards 24
     2.1   Purpose and objectives 24
     2.2   History and overview of European Directives and Standards 25
     2.3   The European Machinery Directive..33
     2.4   Conformity procedures 39
     2.5   Other 'New Approach Directives'..44
     2.6   User side directives: workplace health and safety legislation 46
     2.7   Some machinery safety standards...49
     2.8   Regulations and standards in the USA 52
     2.9   Conclusions...55
            References...55

3   Risk assessment and risk reduction 56
     3.1   Purpose and objectives 56
     3.2   Introduction to risk assessment 56
     3.3   Procedure for risk assessment 57
     3.4   Hazard study methods...66
     3.5   Risk estimation 71
     3.6   Risk reduction principles...79
     3.7   Outcomes of the risk assessment..85
     3.8   Documentation methods for the risk assessment 90
     3.9   Conclusions..91
            References..91

4		Design procedures for safety controls	92
	4.1	Introduction to design techniques	92
	4.2	Review of design standard EN 954-1	93
	4.3	Procedure for the design of safety controls based on EN 954	94
	4.4	Design considerations	97
	4.5	Safety categories	104
	4.6	Conclusions	110
		References	111
5		Emergency-stop monitoring and the safety relay	112
	5.1	Introduction	112
	5.2	Definitions and implications of stop functions	112
	5.3	Safety relay terminology	114
	5.4	How does an E-stop safety relay work?	116
	5.5	Practical safety relays	117
	5.6	Certification	125
	5.7	Functional overview of monitoring relays	125
	5.8	Electronic and programmable E-stop monitors	127
	5.9	Using monitoring safety relays for guards (safety gate monitors)	128
	5.10	Review of other monitoring relay functions	128
	5.11	Conclusions	130
		References	131
6		Sensors and devices for machinery protection	132
	6.1	Contents summary	132
	6.2	Purpose and objectives	132
	6.3	Review of guards	138
	6.4	Sensing devices for guards	144
	6.5	Mechanical trapped key interlocking	152
	6.6	Presence sensing devices	154
	6.7	Control devices for safety	163
	6.8	Safety networks and sensors	166
	6.9	Conclusions	168
7		Application guidelines for protection devices	169
	7.1	Introduction	169
	7.2	Choosing protection methods	170
	7.3	Guarding devices	171
	7.4	Point of operation devices	173
	7.5	Application guidance notes for light curtains	180
	7.6	Conclusions	188
8		Programmable systems for safety controls	190
	8.1	Introduction	190
	8.2	Benefits and disadvantages of safety PLCs	195
	8.3	Characteristics of safety PLCs	201
	8.4	Application software	214

8.5   Safe networking...215
8.6   Classification and certification of safety PLCs 218
8.7   Summary...219
       References...219

9   Introduction to standards for programmable systems 220
9.1   Introduction..220
9.2   Objectives..220
9.3   Outline of IEC 61508 221
9.4   Concept of SILs...226
9.5   How can we determine the required SIL for a safety function?............228
9.6   Some implications of IEC 61508 for machinery systems 230
9.7   Summary..232
9.8   Conclusion..232
       References..233
       Appendix: Notes on the method for the determination of SILs
       for a machinery safety application...234

Appendix A: References and sources of information on machinery safety 240

Appendix B: Glossary 243

Appendix C: Notes on tolerable risk..248

Appendix D: Notes on PUWER 252

Appendix E: Guide to fault tree analysis 257

Practical exercises
       Exercise 1...262
       Exercise 2...263
       Exercise 3...264
       Exercise 4...266
       Exercise 5...268
       Exercise 6...269

Answers to practical exercises
       Exercise 1...270
       Exercise 2...272
       Exercise 3...274
       Exercise 4...277
       Exercise 5...279
       Exercise 6...281

Index 283

Preface

The technology of safety-related control systems plays a major role in the provision of safe working conditions throughout industry. Regulations require that suppliers and users of machines in all forms from simple tools to automated manufacturing lines take all necessary steps to protect workers from injury due to the hazards of using machines. Perhaps your company is wasting money on inappropriate safety measures that still do not deliver compliance with local safety regulations? This book aims to provide you with the knowledge to tackle machinery safety control problems at a basic and practical level whilst following the best available international standards. The book begins with an overview of machinery safety issues, introducing the concepts of hazard identification and risk reduction. The major international standards that are used to support compliance with EC regulations are highlighted and these standards are used as a basis for the design procedures. This approach will assist you to follow best practices for safety system applications wherever your plant is situated. The book looks at the risk assessment processes used to identify hazards and to quantify the risks inherent in a machine. This enables engineers to evaluate the need for risk reduction and hence define the safety functions to be provided by safety-related electrical controls. The book then introduces the concepts of safety categories as defined by standard EN 954 and illustrates the principles of failsafe design, fault tolerance and self-testing. With design procedures established the book now provides an introduction to machinery protection devices such as guards, enclosures with interlocks and guard monitoring relays, locking systems, safety mats, photo electric and electro sensitive principles and the application of light curtains.

The book continues with a study of Safety Control System techniques and introduces the principles of safety-certified PLCs focussing on practical useful information. Application examples such as guard door interlocking applications, two-hand controls, muting, area protection of robot installations and motion detection are then discussed.

The recently established standard IEC 61508 for functional safety of programmable systems is outlined. The concepts of safety integrity levels (SILs) are briefly explained and the key issues associated with software based safety applications are highlighted.

Typical people who will find this book useful include:

- Instrumentation and control engineers and technicians
- Process control engineers and technicians
- Electrical engineers
- Consulting engineers
- Process development engineers
- Design engineers
- Control systems sales engineers
- Maintenance supervisors
- Compliance engineers
- Machinery designers and system integrators
- Safety professionals, health and safety officers
- Production managers

- Automation engineers
- Test engineers.

We would hope that you will be able to do the following as a result of reading this book:

- Identify hazards that occur with machinery and make them safe
- Describe the typical and widely used regulations for Safe Machinery use
- Apply the design procedures for Safety Controls
- Understand the Regulations that apply to manufacturers and users of equipment
- Apply safety rules to your next design involving guards, electrical and safety systems
- Perform simple risk assessment and hazard study methods to your project
- Understand machinery protection devices
- Know when to use Safety PLCs and how to apply them effectively
- Apply basic principles of Machinery Safety Management.

A basic working knowledge of electrical engineering concepts is useful but not essential as there will be a brief revision at the commencement of the class.

1

Introduction to the machinery safety workshop

The safety of machinery affects all of us in everyday life, at home or at work or at leisure. Machines are part of our lives and our safety is dependent on the machines being safe for us to use at all times. So how should a machine be made safe? There are some very basic aspects of safety that spring to mind. A machine should be:

- *Physically safe:* No sharp edges, spikes or projections we can bump into. No chance of it falling over onto somebody. No ways in which it can throw objects around or let out jets of steam or noxious gases. No chance of explosions or radiation.
- *Mechanically safe:* The moving parts must not be able to hurt someone. If there's a risk that this can happen then we need protection measures: fixed guards, movable guards, area-sensing devices that stop the machine quickly if someone is in the danger zone.
- *Electrically safe:* There must be no chance of an electrical shock or a dangerous electrical circuit arrangement.
- *Functionally safe:* All the stop switches, guards and safety-sensing devices that may be there to protect us must function properly. All safety controls that prevent movement at the wrong time must be reliable.

This workshop concentrates mainly on functional safety systems, those safety measures that are based on sensors and control systems that are designed to ensure safe working of the machines. These are also known as safety-related electrical control systems (sometimes abbreviated as SRECS). The workshop training is intended for technicians and development engineers who will be concerned with designing and maintaining safety-related control systems for automated machinery.

We shall also be looking at the general requirements for safety of machines, including some aspects of mechanical guarding and electrical equipment safety.

As with all safety system applications, the technical requirements must be supported by a basic understanding of risk management principles. These principles provide guidance on the extent and complexity of essential safety measures for each application. Once a safety system has been devised, its success depends on both the technical quality of the design and the effective management of all aspects of the safety system throughout its life cycle. This workshop therefore combines basic training in the principles of safety management, with specialized chapters on the safety devices and techniques commonly seen in industry.

We shall see that there is a common approach to most safety applications involving electrical/electronic control systems. If we can identify the ground rules and the common features that apply to most safety applications in machinery, we shall have a basis or framework for tackling any particular project.

This is the basis of our workshop:

- Identify the common factors in most machinery safety applications.
- Outline the framework of regulations and standards that support good safety practices.
- Develop a basic knowledge of design principles and design practices.
- Develop a procedure for defining safety requirements and for selecting appropriate safety devices.
- Learn about the most widely used safety techniques and see how they are used in practice.
- Introduce the current and newly developing technologies for safety systems.

At the end of the workshop we hope that you will have sufficient knowledge to approach any machinery safety project or maintenance situation with confidence. You should feel that you have the background training to recognize the basic features of safety systems and to know the principles on which they should be built.

1.1   Scope and objectives

This chapter provides an introduction to some key topics in machinery safety. The topics include:

- The definition of a machine and its safety-related controls
- Regulations and standards
- Hazards and risk assessment
- Concepts of risk reduction and tolerable risk
- An introduction to the safety life cycle and its relevance to safety management
- A simple example of a machine safety system and its development steps
- Safety equipment, sensors, logic solvers and actuators
- Standards for programmable systems
- Application of safety programmable logic controllers (PLCs) and bus networks.

The topics will be studied in more detail in the following chapters but the objective here is to achieve the broadest possible view of the subject before diving into particular details.

1.2   Machinery and controls

What do we mean by machinery?

As you might expect, almost any assembly of mechanical and electrical equipments which has moving parts can be considered a machine. Various definitions of machines are offered in engineering standards.

This definition of machinery is taken from the European standard EN 292-1: *Safety of machinery – Basic concepts, general principles for design.*

Machinery (machine)
An assembly of linked parts or components, at least one of which moves, with the appropriate machine actuators, control and power circuits, etc., joined together for a specific application, in particular for the processing, treatment, moving or packaging of a material.

The term machinery also covers an assembly of machines, which, in order to achieve a common function or deliver a product, are arranged and controlled so that they function as an integral whole.

The electrical safety standard IEC 60204-1 adds the following detail (in paragraph 3.33):

Machinery also means interchangeable equipment modifying the function of a machine, which is placed on the market (supplied) for the purpose of being assembled with a machine or a series of different machines, or with a tractor by the operator himself insofar as this equipment is not a spare part or a tool.

It can be seen that this definition will embrace a vast range of equipments. Typically we are interested in familiar types of machinery and there are some obvious groupings:

- Domestic applicances
- Lifts and escalators, cranes and hoists, forklift trucks
- Basic cutting, sawing and drilling tools
- Machine tools such as lathes, milling machines, metal working drills, circular saws
- Press tools ranging from small ones for components to large presses for motor vehicle body parts
- Multi-station machining centers
- Assembly lines and conveyor systems where multiple machines are coordinated to provide a complete manufacturing process
- Robots and robot-operated assembly or packing units
- Agricultural machines such as combined harvesters and baling machines.

In all the above machines it is the responsibility of the builder and supplier to ensure that the machine is designed to be safe to use in its intended manner. This very often requires that the machine be fitted with essential safety measures to minimize the risk of injury to people near to the machines, particularly those operating and maintaining the machines.

What is a machinery safety system?

Any assembly of devices designed to protect people from hazards or injuries that could arise from the use of the machine can be considered to be a *machinery safety system*. The machinery safety system may also provide protection for the machine itself or other machines against damage due to malfunctioning of the machine. Let us look at a simple diagram of a machine with its basic control system (see Figure 1.1) and then see where the safety system fits in.

Figure 1.1 depicts a machine with a basic control system. It may, for example, have drives creating movements of assemblies and cutting tools; if it is an injection-molding machine it may have hydraulic pumps with hydraulic valves controlling linear actuators. The actions of the machine will have physical parameters that can be measured with sensors and evaluated by the control system. The control system will operate drives and actuators to follow a program of actions that will be decided by the operator and/or the stored program within the machine.

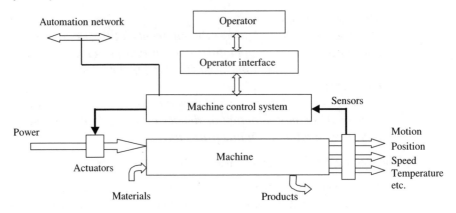

Machine with basic control system

No safety-related parts are identified in this diagram

Figure 1.1
Block diagram model of a typical machine

In automation systems it may be that the machine controls will exchange data with a larger control network, enabling this machine to be operated in coordination with several other machines. Hence we must recognize that there are several sources of commands for the machine to respond with controlled actions. Sources of commands are:

- The operator via a control interface
- The machine control logic from a fixed logic control or from a stored program
- The automation cell control system.

To these we must add 'false commands' from malfunctions:

- The machine goes wrong, mechanically or electrically
- The operator does something wrong
- The control system goes wrong or is incorrectly programed.

Any of these commands could cause the machine to start moving and hence there is a possible hazard if a person or another machine is in the wrong place at the time.

Fixed guards are usually the first line of defense to prevent a person being hurt by the machine but in many cases the situation will require a logical action from the control system to prevent movement or other physical events from happening until safe conditions are proved to exist. These protective measures are the 'safety functions' to be provided by the control system. Those parts of the basic control system as well as any specially provided safety parts are known as the 'safety-related parts of the control system'. In Figure 1.2 they are shown to consist of safety critical parts of the basic controls (e.g. emergency-stop controls) as well as separate sensors for devices such as the presence-sensing light curtains or safety mats.

It is important to bear in mind that the safety-related controls include all parts involved in the safety function. Hence the sensors, logic or evaluation units and the final drive interlocks and contactors or valves belong to the safety control system.

Whilst some safety devices can simply be passive guards such as shields or covers, it is most likely that many of the safety functions will be provided by a combination of mechanical devices and an SRECS. The elements of an SRECS are as shown in Figure 1.3 and it is worth noting that these are very similar to those required for a process SIS.

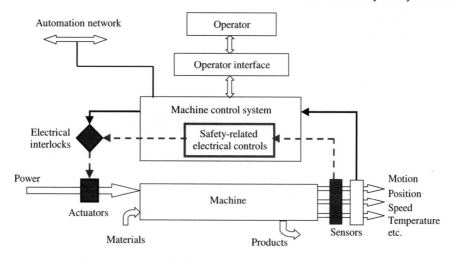

Machine with safety-releated electrical parts

Any part of the machine control system that deals with safety becomes a safety related part

Safety-releated electrical control systems exist within the machine control system but
operate independently of all other functions

Figure 1.2
Block diagram of machine showing safety-related parts

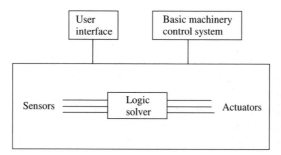

Figure 1.3
Basic elements of a safety-related control system

Figure 1.3 depicts the essential elements of all safety-related control systems. These comprise:

- The safety control equipment comprising sensors, logic solvers and actuators.
- An interface to the basic control system that must not allow the basic controls or operator settings to interfere with or corrupt the safety function.
- An interface to the users; these will be operators, machine setters, technicians, engineers. This interface must also be secure against corruption of the safety function.
- Functional separation: We want to keep the safety systems functionally independent from the basic controls to protect them against being accidentally or deliberately defeated by action of the basic controls.
- Avoidance of common cause failures. We want to avoid the possibility that a malfunction or electrical defect in the basic machinery controls can at the same time override or corrupt the safety controls. For example, if one PLC output stage controlled the starter for a drive and also controlled a safety interlock it would be useless as a safety device if the PLC failed with all outputs on.

Figure 1.4 represents a very simple safety control scheme typically required for a machine tool to protect operators against getting entangled in rotating parts.

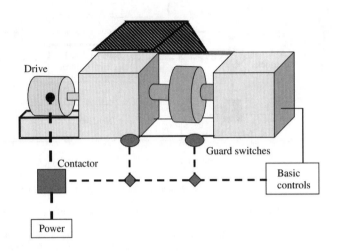

Figure 1.4
Elementary guard position interlock with guard open, drive stopped

The interlocks prevent the spindle drive from starting unless the guard is closed. Failure of any part of this interlock system increases the risk of an accident. It is easy in this example to see that the limit switches and final contactor form part of the safety function.

A typical hardware-based implementation of the guard door safety function will link the guard door switches in series with an E-stop switch to provide an input to a latching relay. The latching relay will trip when the guard door is opened or when the E-stop is pressed. To improve the safety of the circuits an additional relay is used to prevent the latching relay from being reset unless the safety control circuits are healthy (i.e. free of dangerous faults). For example, in Figure 1.5 a simplified safety relay design is shown where K3 is a relay that must be energized before the latching relay K1 can be set. K3 will not energize unless the power control contactor(s) C has been released, proving that it is not held in by another stray circuit or by a mechanical defect.

In practice, relay K1 is usually duplicated by a second channel or redundant relay K2 and both relays must be energized and latched to close the output circuits. K3 is often arranged with multiple contacts and expansion units to enable many drives to be interlocked from the same logic.

The example shown in Figure 1.5 uses a safety-monitoring relay unit to perform the essential logic functions required to provide safety integrity. These are: checks on the state of input signals, detection of stuck contactors, wiring faults in the input and output circuits, timing and logic for interlocking control, etc. The safety-monitoring relay modules ensure that the safety interlocks and E-stop functions are able to operate independently of the basic control system actions at all times.

These are some of the key design features we shall be keeping in mind throughout the workshop. Later in the workshop we shall be looking at ways of achieving functional independence for the safety systems whilst achieving the cost and performance benefits of a physically integrated control and safety system.

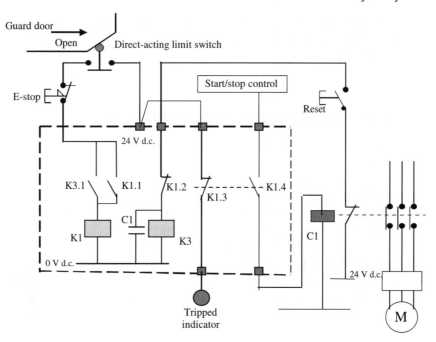

Figure 1.5
Simplified circuit of an E-stop and guard-monitoring relay

1.3 Distinction between machinery and process safety control systems

There are important parallels between process safety systems and machinery safety. These are worth noting because many technicians and engineers will have to deal with safety systems in both categories. There is also an increasing trend to share the technical standards across these industries, and some vendors offer safety equipment that is suitable for both.

For process technology the identification of unacceptable risks leads to a set of risk reduction measures that often include what is known as a safety-instrumented system (SIS) or emergency shutdown system.

- Process plant shutdown systems define the grade or performance of their applications in terms of safety integrity levels (SILs).
- Machinery safety systems are traditionally defined for performance by 'safety categories' but will in future be moving to the same basis of SILs for complex and/or programmable safety systems.

Process plant safety is subject to different regulations and design standards from those applicable to machinery safety but the basic principles are essentially the same.

Some interesting questions arise when a section of process plant has a large and dangerous machine in the plant.

- Is the hazard coming from the process or from the machine?
- Which regulations are applicable?
- What design standard shall we apply?

If the hazard is due to the process, the plant safety systems can deal with it. If the machine presents hazards of its own, the safety requirements will fall under machinery safety regulations.

1.4   International standards and practices

It is a characteristic of safety legislation in most industrialized countries to have an overall requirement for safety at work in the form of general occupational health and safety regulations. The regulations then refer to a subset of regulations directed at particular aspects of hazards at work.

Regulations such as OHSA (in USA) and ESHWR (in Europe) require all companies to ensure the safety of workers, environment and plant. Safety practice begins in all industries with the practice of risk assessment, requiring companies to identify and evaluate risk in the workplace and to record the measures, if any, that they have taken to minimize the risks. When it comes to the provision of measures to improve safety there is a difference in approach for some industries, as noted below.

1.4.1   Safety engineering methods in process plants

In the case of process plants such as refineries or chemical works the laws leave the details of the engineered safety systems to be satisfied by a set of widely applied 'best practices' to be used at the discretion of the end user. There are some standards (such as IEC 61511) that set down the principles for management and design of the safety systems for process plants. The owner or operating company is then obliged to justify the details of the safety measures for each application.

1.4.2   Safety engineering methods in machinery

In machinery applications, the legislation either prescribes adherence to a set of named standards (USA practice) or allows us to presume compliance if we follow the relevant standards (EU practice). If there are no standards applicable for a particular machine, a safety case can be established by using the general principles applied in all safety applications. These principles are set down in higher level, general-purpose standards. It is these high-level standards that are of particular interest to us in this workshop since they provide a good basis for essential training in the subject.

1.4.3   International standards

Both process and machinery safety methods are part of a growing trend to set common standards for safety practices that will be acceptable all across the world. Figure 1.6 shows some of the major regulations and standards that have become established in Europe and in the USA.

Control of Major Accident Hazards regulations (COMAH) is a European Union (EU) requirement for managing safety in large hazardous processes. Similar requirements exist in the USA under the process safety rules of OHS regulations (OSH 29 CFR 1910.119).

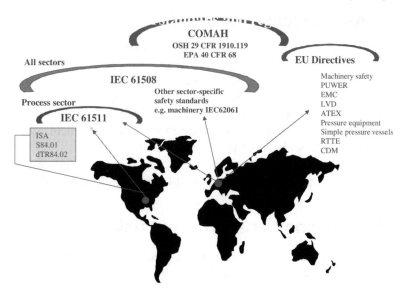

Figure 1.6
Legislation and standards for machinery and process safety

Control of Substances Hazardous to Health regulations (COSHH) is a UK regulation to ensure that any factory handling or processing hazardous substances takes steps to minimize the risk of substances harming people or the environment. This is similar to the USA's clean air act requirements of the Environmental Protection Agency (EPA) (EPA 40 CFR 68).

The EU Machinery Directive defines machinery safety requirements to be observed in EU states by manufacturers, suppliers and users of machines. It references a wide range of general and detailed engineering standards that have been 'harmonized'. This means that they have been accepted by each of the member states as a national standard in that country. The great value of working to the requirements of a harmonized standard is that it creates a 'presumption of compliance' with the relevant EC Directive. This simplifies the task of proving that the machine will meet the requirements of the Safety Directive.

There are other EU Directives that impact on machinery equipment such as the Low Voltage Directive (LVD) and we shall look more closely at this in Chapter 3.

In the United States there is general intention to achieve uniformity with European standards so that there can be free interchange of products and services. The OHSA regulations incorporate and require compliance with the *ANSI B 11 series of standards* produced by the Association of Manufacturing Technology (AMT), a trade association of the machine tool industry.

1.4.4 Supplier's responsibility for safety

An important point to note about machinery safety legislation is that the designer and builder of a machine has a major responsibility to make the machine safe to use within a foreseeable range of applications. Since the machine may find its way into a wide variety of workplaces and into domestic homes in the case of home appliances, safety must be built into the machine as a unit. This makes the supplier of the machine responsible for proving it is safe to use. The supplier can be prosecuted for supplying an unsafe machine.

1.4.5   Owner's responsibility for safety

Once the machine is installed in a factory it becomes the owner's responsibility to see that it is used in a safe manner and that all safety measures are properly maintained and applied. The owner will, of course, want to buy a machine that comes with all the safety measures in place. However as soon as two or more machinery devices are assembled to form a production unit, the user has created a new and often unique machine. Hence there will always be a need for the user to do risk assessment and to implement additional safety measures whenever the need is found.

It follows that both the suppliers and the end users should have a good knowledge of the range of applicable regulations and their supporting standards.

1.5   Introduction to hazards and risks

The first step in any safety-related project is to identify the hazards and to consider the level of the risks they present. *So what are hazards and what is risk?*

1.5.1   Hazard

In the broadest terms, a *hazard* is an inherent physical or chemical characteristic that has the potential for causing harm to people, property or the environment. In machinery usage EN 292-1 describes Hazard as '*A source of possible injury to damage to health*' and it goes on to describe some elementary forms of mechanical hazard in the following list of hazard types:

- Crushing
- Cutting or severing
- Entanglement
- Impact
- Stabbing or puncture
- Friction or abrasion
- High-pressure fluid ejection.

Other types of hazard may also be present such as the primary chemical process hazards:

- Explosion
- Fire
- Toxic release.

And we have already mentioned electrical hazards. The first task of any risk assessment is to identify the potential hazards of a machine and then move on to evaluate the level of risk they present.

1.5.2   Risk

Risk is usually defined as the combination of the severity and probability of an event. In other words, how often can it happen and how bad is it when it does happen? Risk can be evaluated qualitatively or quantitatively. Roughly

$$\text{Risk} = \text{Frequency of the event} \times \text{consequence of hazard}$$

In EN 1050 a simple diagram similar to Figure 1.7 describes the elements of risk.

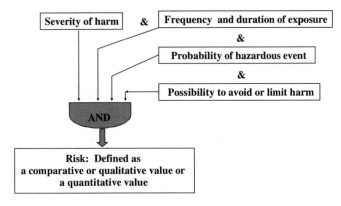

Figure 1.7
Elements of risk are combined to produce a qualitative or quantitative value

Qualitative descriptions of risk use terms such as 'low', or 'high' or 'severe'. *Quantitative descriptions* of risk use numerical values such as '1 irreversible injury per 1000 years'; this might be the equivalent of a 'medium but unacceptable risk'. If the quantitative risk is reduced to say '1 irreversible injury per 100 000 years' we might describe this as a 'low and acceptable risk'.

1.6 Risk reduction

The reduction of risk can sometimes be achieved by design improvements but if this is not practicable it often requires protection measures. In some cases this will be an alternative way of doing things or it can be a protection system such as an SRECS. The design principle is shown in Figure 1.8.

As the figure shows, we have to evaluate the risks due to the hazards and then compare them with the target risk levels. To design a protection system we have to specify what safety function it has to perform and then define how good it must be (define the safety integrity).

The objective is to reduce the risk from the *unacceptable* to at least the *tolerable*. This seems simple enough as long as we can work out what is tolerable. See Figure 1.9 for an everyday example of risk reduction and tolerable risk principles applied on the cricket field.

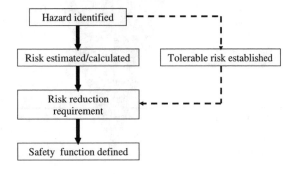

Figure 1.8
Risk reduction steps

Risk reduction example: the fast bowler

If we can't take away the hazard we shall have to reduce the risk.

This means: reduce the frequency and/or reduce the consequence

> Example:
> Glen McGrath is the bowler: His bouncer is the Hazard
> You are the batsman: You are at risk
> Frequency = 6 times per over. Consequence = Ouch!
>
> Risk = 6 × Ouch !
>
> Risk reduction: Limit bouncers to 2 per over. Wear more pads.
>
> Risk = 2 × Ouch !

Figure 1.9
Sporting example for risk reduction

Safety systems are all about risk reduction. If we can't take away the hazard we shall have to reduce the risk.

Risk reduction can be achieved by reducing either the frequency of a hazardous event or its consequences or by reducing both. Generally the most desirable approach is to first reduce the frequency since all events are likely to have cost implications even without dire consequences. So for a typical problem of physical harm from moving parts of a machine the risk reduction is achieved by reducing the possibilities that a person can get in the way of the moving parts. If we can reduce the chance of trapping a hand in the moving parts from say once per week to perhaps once per hundred years we may feel that this is an acceptable solution. In this case we have settled for what is known as a *tolerable risk*.

1.7   The Alarp principle for tolerable risk

Figure 1.10 illustrates the concept of tolerable risk and is known as the as low as the reasonably practicable (Alarp) diagram.

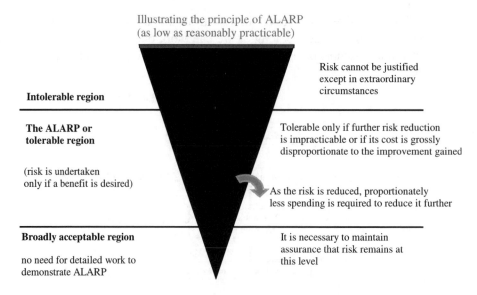

Figure 1.10
Typical Alarp diagram

The Alarp principle recognizes that there are three broad categories of risks:

1. *Negligible risk:* Broadly accepted by most people as they go about their everyday lives, these would include the risk of being struck by lightning or of having brake failure in a car.
2. *Tolerable risk:* We would rather not have the risk but it is tolerable in view of the benefits obtained by accepting it. The cost in inconvenience or in money is balanced against the scale of risk and a compromise is accepted. This would apply to traveling in a car, we accept that accidents happen but we do our best to minimize our chances of disaster. Does it apply to bungee jumping?
3. *Unacceptable risk:* The risk level is so high that we are not prepared to tolerate it. The losses far outweigh any possible benefits in the situation.

Essentially this principle guides the design engineer and the safety specialist into setting tolerable risk targets for a hazardous situation. This is the first step in setting up a standard of performance for any safety system. The problem here is that it is difficult to determine what is a tolerable risk.

Some of the engineering standards simply state that the machine must be 'safe'. If we look in the standards for a definition of safety we get: 'Freedom from unacceptable harm'.

This seems to be the same thing as acceptable risk but doesn't get us any further. We shall take a more detailed look at 'acceptable' or 'tolerable' risk criteria in Chapter 3 as we follow the risk reduction steps described in the standard EN 1050.

1.7.1   Risk assessment procedure

The process for a risk assessment for the handling and use of machines follows the same general rules for all risk assessments. These rules are most clearly described in a widely used brochure published by the UK Health and Safety Executive (HSE) called 'five steps to risk assessment'. We recommend readers to take a free download of this leaflet from the HSE website: www.hse.org. The five steps recommended in the leaflet are shown in Figure 1.11. These simple risk assessment steps define the basis for our work on machinery safety just as they will apply to a wide variety of activities in the workplace.

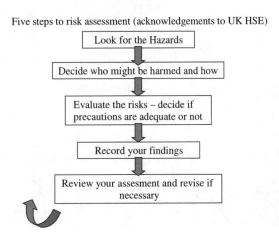

Five steps to risk assessment (acknowledgements to UK HSE)

Figure 1.11
Five steps in the risk assessment procedure

If we decide that the precautions are not adequate it will be clear that certain steps would be taken to improve the situation. Typically these steps are to be based on the following responses given in order of preference:

1 Try a less-risky option
2 Prevent access to the hazard (e.g. by guarding)
3 Organize work to reduce exposure to the hazard
4 Issue personal protective equipment
5 Provide welfare facilities (e.g. washing facilities for removal of contamination and first aid).

In particular, items (2) and (3) above will be relevant to our work on the development of machinery safety systems.

1.8   Development example for a machinery safety system

In Figure 1.12 we can see a typical example of machinery safety practices by looking at a metalworking or woodworking center lathe. One of the most widely used of all machine tools, the center lathe presents some basic hazards. For example:

- The spinning chuck or spindle presents hazards such as entanglement of clothes or possibly abrasions, cuts or bruises to a hand or arm coming into contact with it.
- The cutting of metal can produce flying chips. An impact hazard including damage to eyes.
- An exposed lead screw presents a hazard of entanglement for clothes or trapping of hands.

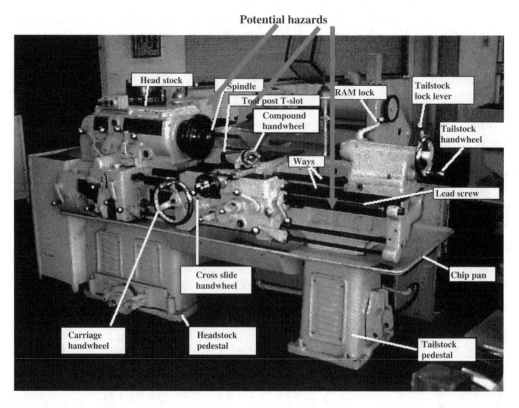

Figure 1.12
Metal working lathe for risk assessment

These three hazards present various levels of risk to the person using the lathe. The machinery safety systems are provided to reduce the risks presented by these hazards to levels that are considered reasonable or tolerable.

1.8.1 Risk assessment example

Here is an elementary risk assessment for the lathe example. The risks might be evaluated as shown before the application of measures to reduce the risks.

Hazard	Probability of Event	Consequence	Risk
Operator contact with spinning chuck	High avg. 1 per week	Abrasion wounds	Abrasion wounds once per week
Flying chips hit face	Very high avg. 1 per day	1 in 10 chances of eye damage	Eye damage once every 10 days
Entanglement of clothes with exposed rotating lead screw	Moderate avg. 1 per year	1 in 5 chances of broken arm	One broken arm per 5 years

Clearly the risks shown in this table are unacceptable and they have to be reduced. Risk reduction options consist of ways of reducing the probability of the event and/or reducing the consequence. In Chapter 3 we study risk assessment methods and ways of deciding what is tolerable.

1.8.2 Propose safety functions

For the moment if we assume that the risks have to be reduced it is easy to see that some typical safety measures can be applied. For example:

- The exposed lead screw can be made safer by a telescopic or flexible cover that remains in place at all times except when the machine is stripped for service. This is a mechanical guard that normally has no requirement for interlocking to the electrical drives.
- Where there is danger from flying chips it may be acceptable to wear protection equipment (usually abbreviated: PPE); in this case, safety glasses.
- A lathe guard can be provided to cover the spinning chuck. In the slides shown here a simple hinged cover can be mounted to be put in place by the operator after he/she has set up the workpiece and tightened the chuck jaws.

But now we have to be sure that the operator always swings or slides the cover into position. We want to be sure that the lathe cannot be operated if the cover is out of position.

This means we shall want to arrange an electrical interlock to make sure that the lathe will not start turning until the guard is in place. To do this we need to have a position-sensing switch, perhaps a mechanical limit switch, set up to ensure the guard is in position before it will close its contact. Figure 1.13 shows the guard in the open position with power to the drive interrupted by the limit switches. Figure 1.14 shows that power is enabled when the guard is closed over the hazard. Here we have the beginnings of an SRECS. This particular safety function requires that electrical power to the lathe drive will be switched off if the guard is not in position.

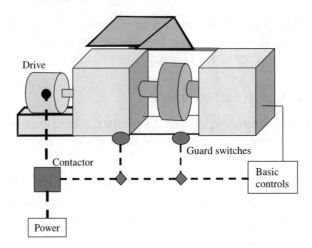

Figure 1.13
Elementary guard position interlock with guard open, drive stopped

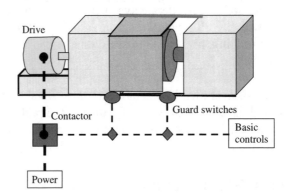

Figure 1.14
Elementary guard position interlock with guard closed, drive can be started

1.8.3   Risk assessment after adding protection measures

The table we saw at the start of this exercise can now be updated to show the effect of protection measures. This is a typical risk assessment reporting method.

Hazard	Probability of Event	Consequence	Risk Before	Safety Measure	Risk After
Operator contact with spinning chuck	High	Abrasion wounds	High	Interlocked guard	Low
Flying chips hit face	Very high	1 in 10 chances of eye damage	Severe	Interlocked guard and PPE goggles	Very low
Entanglement of clothes with exposed rotating lead screw	Moderate	1 in 5 chances of broken arm	Medium	Flexible cover	Very low

1.8.4   Evaluate expected risk reduction

It looks as if the safety interlock and guards we have specified will do the job very well. If we follow the risk reduction procedures, what we need to do now is check to see if the new level of risk is acceptable or tolerable. This seems simple enough at first. But to be sure that we have got it right we have to consider possible problems due to failures of the equipment or due to incorrect design. This takes us into the subject of 'safety integrity' and how it can be determined. We shall look at the whole subject of failure modes, reliability analysis and safety integrity at relevant points throughout the workshop.

Consider failure modes and limitations of the protection measures

Let us look again at the lathe guard example. What could go wrong? What are the chances? No safety device can achieve 100% reliability. For example:

- The limit switch must be good enough to always do its job even when the guard gets a bit worn and doesn't locate so well. So it has to have a good range of tolerance for positioning errors.
- We don't want someone to jam a matchstick into the switch so that the guard function can be defeated. So it must be tamper-proof.
- If the cover is lifted or moved away whilst the chuck is spinning, the rundown time may not be fast enough to avoid an accident. So may be the cover should be locked in place until the chuck has stopped. This will require some timing or speed-sensing device and an electronic lock. Is this expense and complexity justified? How do we decide?
- If the limit switch does develop a fault we want to be sure that the safety of the guard function is not lost. So it should be fail-safe or it should be able to carry on protecting us even when it has a fault (fault-tolerant). Better still we would like to know about the fault as soon as it develops. We may want the safety system to be self-testing (also known as having diagnostics).

The guard and its sensing system have to be designed such that it will not be an obstacle to high productivity. It must not get in the way of efficient use of the machine. It must not present temptations or incentives for people to do without it (bypassing).

The cost of the equipment must not be so high that users are heavily penalized for ensuring safety.

Similar possible problems arise with the circuits and relays or programmable controllers that may be used in linking the limit switch to the drive interlock. Finally we have to make sure that the power break contacts to the drive control cannot be defeated either by a fault or by the actions of another control system or even by the maintenance technician.

So it is the designer's responsibility to see that the safety devices are fit for purpose and it is the maintenance technician's job to keep the devices in good working condition. Both parties must understand the design principles and safety functions of the devices.

1.8.5   Equipment choices for the safety systems

The workshop will examine some of the features of the equipment and devices available to us. We must be able to recognize the benefits and any weaknesses of our equipment choices. In particular, the choices must balance safety performance, capital cost and the effect on productivity.

1.8.6   Standard solutions to standard problems

In many practical projects, the writers of machinery safety standards and the suppliers of components have done a lot of the design job for us for the most common types of machines and for most applications. So our job is to find out what's out there and how to make the best use of it. We get a lot of help from the industry specialists.

- Standards such as EN 954 define safety categories suitable for graded levels of risk reduction service.
- Manufacturers offer safety products designed specifically for the most widely needed safety functions.
- Testing authorities certify that safety devices are fit for the designated tasks and certify the safety category that can be achieved.

The following are some of the electrical and electronic control equipments available in the market for machine safety, arranged in approximate categories:

- Emergency-stop switches
- Safety gate position limit switches, tongue or cam-operated
- Monitoring safety relays for:
    - E-stops
    - Guard positions
    - Two-hand controls
    - Speed monitors and timers
- Muting systems
- Locking safety switches, interlocking devices, trapped key systems
- Electro-sensitive presence sensing devices including:
    - Edge sensing
    - Safety mats
    - Safety light screens/curtains
- PLCs for safety applications
- Certified software applications for commonly used safety functions
- Bus networking of sensors and logic controllers for complex safety applications.

In the workshop we shall be looking at the principles of the different protection methods and will hope to see the factors that will help us to make the best choice for any application.

1.8.7   Programmable systems for automation safety

Programmable systems have become established in machinery safety and there are many new developments taking place at the high-tech end of the market. We shall take a look at the technologies later in the workshop. For the moment we can just list some of the reasons why we would want to use programmable electronics and networks in safety systems:

- Sequencing of shutdown actions in large machines.
- A manufacturing line consisting of several closely linked machines will have many safety functions. For efficiency they need to be implemented under one centralized logic system with efficient monitoring and fault detection.

- Software-driven safety functions provide powerful logic tools with flexibility for coping with changing automation functions.
- Efficient diagnostics software speeds up troubleshooting and reduces downtime.
- Networked input/output systems simplify cabling and reduce installation costs.
- Selective shutdown facilities reduce the impact of safety trips on the rest of the plant.
- Cost benefits of re-using software for multiple copies of the same machines.

As stated earlier, the publication of IEC 61508 and the development in progress of a machinery sector version of this standard has set down a firm basis for the use of programmable systems. We shall outline this standard later in the workshop.

1.8.8 Development of integrated safety systems

We have seen that there is a need for the safety systems to be functionally independent of the basic machine control systems. However from the manufacturing point of view there are cost penalties in having to build two control systems for each machine. If you are making, say, several hundred injection-molding machines it would be better if one complete control system could handle both safety and basic control. If you could place all the regular sensors and the safety sensors on one bus network feeding one control box this would be even better.

To a large extent this approach is now becoming feasible without breaking the rules of functional independence and without any loss of safety integrity. Some of the reasons why this approach is gaining ground are:

- Safety PLCs can be made with internal separation of safety and non-safety sections
- Bus systems can achieve safety-rated performance for all sensors
- Continuous diagnostics can ensure fail-safe behavior
- Safety-certified software function blocks can operate in secure partitions of the PLC operating system.

We shall take a brief look at this technology after covering the basics of programmable safety systems.

1.9 The engineering tasks

1.9.1 Introduction to the safety life cycle

The safety products are a great help but they do not relieve the applications engineer of the duty to see that the complete safety function has been designed to meet the original objective. It is always necessary to examine the complete design to see that it meets all aspects of the required safety function and satisfies the required safety category or risk reduction capability. The key to managing this task properly is to plan and execute what is known as the 'safety life cycle' (see Figure 1.15). This simply means all the phased activities from the beginning of the design to the day that someone disposes of the machine.

Here is a rough and informal description of the main steps of design for the safety-related parts of the control system. Let us recap the thinking process we have just been through for the center lathe protection, so we can identify some of the steps in that application.

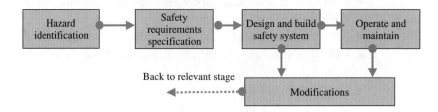

Introducing the safety life cycle

The foundation for all procedural guidelines in safety instrumented systems is the safety life cycle(SLC)

The SLC spans all project phases and has return loops whenever modifications arise

Figure 1.15
Elements and information flow in the safety life cycle

Step 1: Obtain information
Obtain information about the machine and its intended use. (In our case a center lathe used for machining metal objects.)

Step 2: Conduct a hazard identification exercise
For each hazard, analyze the level of risk in terms of consequence of the accident and likelihood of the event. (We listed the hazards, we also decided the consequences and estimated the exposure of the operator and frequency of the possible accident assuming there are no safe guards.)

Step 3: Decide on the measures to be taken to reduce the risk
This involves defining the safety functions to be provided both by design of the machine and by design of safeguarding functions. (We are stuck with the design, so we choose to provide a safety function which will prevent the drive from running unless the guard is in position.) This step includes defining all the essential safeguards such as being tamper-proof.

Step 4: Outline the design of the safety system to identify the subsystems involved in the SRECS
We can see that there will be a position-sensing device with critical requirements and a fail-safe interlocking system to prevent the drive from running, unless the sensor circuit is closed.

Step 5: Specify the equipment and its safety categories
For each subsystem, specify the equipment type you want to use and the level of safety integrity it must have. This is also known as the safety category; the higher the category, the greater the assurance that the subsystem will not fail in a dangerous way. Finding the right category is a matter of knowing the level of risk reduction needed for each application. The standards provide us with further assistance in the way of selection charts. We are referring here to standard EN 954-1 for safety categories. This is a subject we are going to examine in some detail in Chapter 4.

Step 6: Design verification
Carry out a verification check to see that the results we have achieved so far have not been deviated from the original requirements through some misunderstanding or through changes to the original problem analysis. To ensure this is true we need to have a record of all our design work showing how each decision is based on information that is still correct.

Step 7: Detailed design and building
Proceed with detailed design, equipment selection and implementation of the solution. Also define the maintenance and regular testing requirements. Make sure proper test facilities are provided with the equipment.

Step 8: Validation
Check that the design documents and the testing plan are aligned and up to date. Carry out proper testing to demonstrate that the safety functions are fully operational and perform as intended under all foreseeable conditions. Record the results.

Step 9: Provide a design history file or 'technical file'
Describing how the safety system design has been developed, reporting the results of assessments and assumptions, and demonstrating how the design satisfies risk reduction requirements.

Step 10: Use and maintain the safety systems
As intended by the designers, implement a program of regular testing. Keep a record of all tests and enforce strict change control to ensure that the safety system and its design records remain up to date.

The above steps are based on the procedures mapped out in the relevant European Standards but are only an approximate description. We shall look more carefully at the standard procedures in Chapter 3 of the workshop.

1.9.2 Importance of change control

It is one thing to have a well-documented track record for the safety system but the next step in the SLC requires that a procedure be maintained for trapping any changes to the machine that will affect the validity of the present safety design. The basis of change control is that all machinery modifications will be subject to a hazard review against the original hazard analysis documents to see if it has impact on the present safety functions.

If a change is required it must be processed through the relevant steps of the SLC and all updates must be done and recorded properly. This is a hard discipline to follow but it is the best way to maintain or improve the standard of safety that was achieved for the original machine.

Is all of this relevant to maintenance work?

Yes it is. For those involved in maintenance rather than in design it is still important to understand the design processes that should (in theory) lie behind the products you are working with. The latest safety standards require that persons working on safety systems are competent to do so. Competence includes being aware of the design rules and understanding the performance requirements of the safety devices.

It is important for a maintenance technician to fully understand the safety function of the subject equipment and to know the reasons why it has been given a particular safety category. If anything changes in the design of the equipment or in the way the machine is being used the performance requirements of the safety device may be affected. The end user has a continuing responsibility to ensure that adequate safety levels are maintained. This responsibility cannot be properly fulfilled if the reasons for the existing safety measures are not known.

1.10   Benefits of the systematic approach

One of the best advocates for a systematic approach to safety engineering is the UK Health and Safety Executive (HSE): Their publication 'Out of Control' is a very useful little book about 'Why control systems go wrong and how to prevent failure'. The following analysis of 34 accidents attributed to control system failures has been widely published (see Figure 1.16).

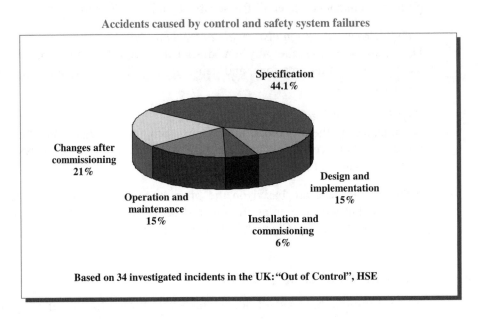

Accidents caused by control and safety system failures

Specification
44.1%

Changes after
commissioning
21%

Design and
implementation
15%

Operation and
maintenance
15%

Installation and
commisioning
6%

Based on 34 investigated incidents in the UK: "Out of Control", HSE

Figure 1.16
Analysis by UK Health and Safety Executive of causes of safety control system failures

The summary by HSE of the problems causing accidents due to control systems includes some useful paragraphs:

The analysis of the incidents shows that the majority were not caused by some subtle failure mode of the control system, but by defects which could have been anticipated if a systematic risk-based approach had been used throughout the life of the system. It is also clear that despite differences in the underlying technology of control systems, the safety principles needed to prevent failure remain the same.

Specification
The analysis shows that a significant percentage of the incidents can be attributed to inadequacies in the specification of the control system. This may have been due either to poor hazard analysis of the equipment under control, or to inadequate assessment of the impact of failure modes of the control system on the specification. Whatever the cause, situations which should have been identified are often missed because a systematic approach had not been used. It is difficult to incorporate the changes required to deal with the late identification of hazards after the design process has begun, and more difficult, (and expensive), to make such changes later in the life of the control system.It is preferable to expend resources eliminating a problem, than to expend resources in dealing with its effects.

Design

Close attention to detail is essential in the design of all safety-related control systems, whether they are simple hard-wired systems, or complex systems implemented by software. It is important that safety analysis techniques are used to ensure that the requirements in the specification are met, and that the foreseeable failure modes of the control system do not compromise that specification. Issues of concern, which have been identified, include an over-optimistic dependence on the safety integrity of single-channel systems, failure to adequately verify software, and poor consideration of human factors. Good design can also eliminate, or at least reduce, the chance of error on the part of the operator or maintenance technician.

Maintenance and modification

The safety integrity of a well-designed system can be severely impaired by inadequate operational procedures for carrying out the maintenance and modification of safety-related systems. Training of staff, inadequate safety analysis, inadequate testing, and inadequate management control of procedures were recurring themes of operational failures.

We can conclude that being systematic:

- Helps us to benefit from previously acquired knowledge and experience
- Minimizes the chances of errors
- Demonstrates to others that we have done the job properly … they recognize our way of doing things is legitimate
- Makes it easier to compare one solution or problem with another and hence leads to generally accepted standards of protection
- Allows continuity between individuals and between different participants in any common venture. Makes the safety system less dependent on any one individual
- Encourages the development of safety products that can be used by many
- Assists suppliers to achieve compliance with regulations.

1.11   Conclusions

This overview has shown us that machines come in all shapes and forms, and they can present us with a number of characteristic hazards. We have seen that there is a systematic method of identifying the hazards and assessing the risks based on the judgment of experienced persons who are needed to estimate the risks, i.e. the likelihood of an accident and the severity of the consequences.

Regulations require that we carry out risk assessments to decide the need for safeguards and the same regulations require that we install and maintain safeguarding equipment to an acceptable design. Engineering standards exist to guide us on what is considered to be acceptable practice both in the design of solutions and in the way we manage the SLC of the machine.

Safeguarding methods range from passive guards to sophisticated safety-related controls but all have the task of reducing risk. The amount of risk reduction needed depends on the original unguarded risk and the perception of what is safe or tolerable. This leads us to the concept that the quality of the solution or the quantity of risk reduction to be applied defines the performance needs of the safety system.

2

Guide to regulations and standards

2.1　Purpose and objectives

The purpose of this chapter is to provide outline information on the typical safety-relevant regulations and legal requirements applicable to both the supplier and the end users of machinery. The information is provided for general guidance only and no responsibility can be accepted by IDC concerning the interpretation of regulations for any country.

We look first at European Union (EU) and United Kingdom requirements because these are where we can find some of the most well-defined requirements. Later sections look at the position in the USA which also has a similar highly developed system of standards and practices. The ground rules for machinery safety are revealed by these examples and will serve as a general guide to the requirements in many other countries.

Contents summary

- Introduction to European Directives, Regulations and Standards
- CE marking and its meaning
- Introduction to the European Machinery Directive
- The EHSWRs of the Machinery Directive
- Routes to conformity for machines
- Obligations of suppliers and users
- Type A, B and C standards
- USA Regulations and Standards
- Concept of Control Reliability.

2.1.1　Objectives

The objectives are:

- Get to know the principles of machinery safety regulations
- Be aware of current EU, UK and USA practices
- Have an outline knowledge of widely used and related standards
- Know the responsibilities of manufacturers, buyers and final users
- Know the steps to be taken to ensure compliance with safety regulations
- Know where to obtain more information on regulatory compliance.

2.1.2   Practices and enforcement by law

The legal aspects of moving machinery safety are generally arranged to cover two primary stages in the life of a machine.

1. *Safe manufacture:* This includes design and installation of plant and equipment
2. *Safe operation:* This includes maintenance and modification of plant and equipment.

Generally, laws and regulations deal with these two aspects separately, although regulators are increasingly aware of the need to improve the links from the supplier to the user. It is reasonable to assume that the UK and USA requirements will serve as a general guide to typical regulations and practices in machinery safety. However, the practices and level of enforcement will vary from country to country. It is one thing to have codes of practice and regulations, and the extent to which these are enforced or observed is another matter. We have to assume that safety practices are essentially good things to do regardless of the level of enforcement. In all cases, the readers are advised to consult their local safety regulators for specific national requirements.

2.2   History and overview of European Directives and Standards

In this section, we try to gain an understanding of how the regulations in Europe are arranged for safety of machines and their use in the workplace. It may be of interest here to look at the evolution of workplace-related safety regulations in UK. The journey began in the nineteenth century:

- 1802   The Health and Morals of Apprentices Act 1802 focused on health and welfare rather than safety. The enforcement was lax and through voluntary unpaid inspectors.
- 1833   The Factory Act empowering government to employ paid factory inspectors. But the act applied only to textile mills and dealt with health and welfare but not safety.
- 1840   Lord Shaftesbury took successful legal action for damages in negligence on behalf of an employee injured in a factory accident (Cotterell vs Stocks). The government attention was drawn toward the fact that major cause of accidents was inadequate fencing and guarding of machinery.
- 1842   The Coal Mines Act contained the first safety provisions.
- 1844   The Factories Act, the three highlights being:

  1. Prohibition on cleaning of machinery that was in motion
  2. Requirement of fencing of certain type of machines
  3. Provided for reporting of accidents.

- 1856   Amendment to act of 1844 to weaken the requirement of fencing.
- Over the years various legislations separately for mines and quarries, agriculture and factories, etc.
- 1901   The Factory and Workshop Act 1901 was the first attempt at consolidating the multiplicity of legislations as regards factories.
- 1937   The Factories Act incorporated major reforms like eliminating distinctions between textile and non-textile factories.

- 1954   Mines and Quarries Act is the principal legislation governing mines and quarries.
- 1961   The Factory Act was another attempt to further consolidate all subsequent legislations. It may be noted that some parts of this act are still operative.
- 1963   The Offices, Shops and Railway Premises Act had consolidated various legislations till that date.
- 1974   Health and Safety at Work Act (HSWA) was a watershed in the history of UK, was all-embracing in its coverage, removing the distinctions between different types of workplace and placing general duties upon employers, employees and others. The emphasis has been on the management of health and safety at work. Some parts of this act are still operative.
- 1986   The Single European Act (SEA) introduces health and safety as an important area of European legislation. Two aspects of SEA are worth noting and are source of all directives issued, which are enforceable through qualified voting majority:

  1. *Article 100A:*   It includes harmonization of technical standards and safety requirements for specific products having implication for health and safety at work through matters such as safety standards for design and construction of machinery. This gave rise to the 'New Approach Directives'.
  2. *Article 118A:*   It states that particular attention shall be paid to encouraging improvements in the working environments as regards the health and safety of workers in order to harmonize conditions within the Union. This gave rise to the 'Framework Directives'.

- 1988   BS 5304 (BSI) Safety of Machinery, no longer a current document but remains a valuable publication on the subject of moving machinery safety.
- 1992   The Supply of Machinery (Safety) Regulations 1992, Statutory Instruments number 3073 of 1992; The Supply of Machinery (Safety) (Amendment) Regulations 1994, Statutory Instruments number 2063 of 1994. These regulations give effect in the UK to the EU Machinery Directive; effectively 'the Supply Law'.
- 1997   The Provision and Use of Work Equipment Regulations (PUWER) giving regulations for the safe construction and use of machinery. This has been amended in 1998 and continues as the primary legislation for workplace safety. These regulations give effect in the UK to the EU Directive for The Use of Work Equipment, effectively 'the User's Law'.

The present-day position is that UK laws are aligned with relevant EU directives whilst incorporating the experience gained from many years of safety legislation. Figure 2.1 summarizes the position into the supply/user framework.

We shall look at the scope and effect of the Machinery Directive next to see what we can find out about the scope and practices of machinery safety.

2.2.1   Directives relevant to machinery safety

Essentially the safety of machinery must be tackled from two sides:

1. The manufacturers and suppliers must build machines that are safe to use.
2. The users of machines (i.e. employers and workers) must ensure the machines are used in a safe manner and the workplace environment is safe for the workers.

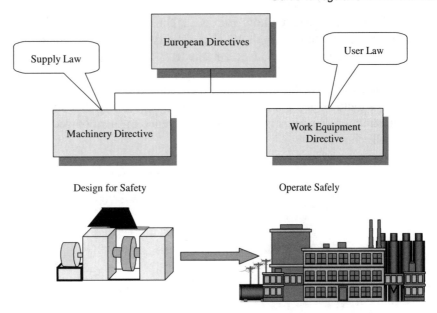

Figure 2.1
EU Directives devolve into laws for suppliers and for users

As seen in Figure 2.1 relevant types of directives exist for both the above groups. The 'new approach' directives define requirements for product safety whilst the health and safety 'framework' directives deal with user's obligations to provide a safe workplace and safe operating conditions.

2.2.2 Relevance of the Directives and Laws for this workshop

Why should we study the laws and directives? Our workshop is looking for practical training, so do we need to know about these laws?

For the purposes of this workshop, it will be helpful to study the scope and effects of the EU Directives and their application in the UK. The applications will help us to understand the principles of machinery safety. We should also bear in mind that most of the technical solutions and devices offered by suppliers of safeguarding equipment have been developed to satisfy legal requirements and have been designed to comply with guidelines provided by the engineering standards linked to the EU regulations.

If you are involved in preparing machinery for delivery to any EU member state, the supply side directives will apply directly to your work, wherever you are in the world.

2.2.3 Principles of EU Directives

The sources of EU law through which the EU regulations are implemented can be divided into three categories:

1. *Primary sources:* Comprising the founding treaties, Community Acts (such as SEA) and further treaties (such as Maastricht or accession treaties).
2. *Secondary sources:* Comprising of regulations, directives and decisions. Through this, EU implements the policy in more detail.
3. *Non-legally binding:* Sources-opinions and other non-treaty acts (such as guidelines, resolutions, communications, etc.).

Figure 2.2 illustrates how European Directives are adopted and implemented into the national legislation by each member country of the EC. Each country may create different laws or adapt existing laws but the end result should still comply with the essential requirements of the EU Directives. The typical result, e.g. in the case of machinery manufacture, will be that machines meeting the safety requirements of one EU member state should be acceptable for use in any other EU member state. Uniformity of design and construction standards is assisted by reference to 'harmonized standards' that are accepted by all member states.

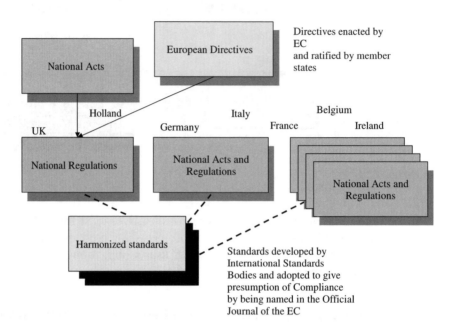

Figure 2.2
EU Directives unify safety regulations

2.2.4   Supply side laws: the EU 'New Approach Directives'

The term 'New Approach Directives' applies to a range of EU Directives that have certain basic principles in common. These include:

- Mandatory essential provisions which apply to the product.
- Requirements for member states to ensure that products not in conformity with essential provisions are not allowed to circulate within the member states.
- The manufacturer is provided with the opportunity to certify conformity with the relevant directives. This leads to the manufacturer placing the CE mark on his product if it is claimed to conform.
- Legislation no longer specifies that specific standards have to be met. However, it can be 'reasonably assumed' that when harmonized standards are met, the associated goals of the EU Directives are fulfilled.
- Manufacturer achieves conformity by compliance with a national law or regulation. This implies compliance with the equivalent laws in any other member state.

The CE mark is the characteristic symbol shown in Figure 2.3 that is applied to a product when the manufacturer or an appointed body certifies that the product conforms to the requirements of all applicable EU Directives.

The CE mark

Minimum height 5 mm

Figure 2.3
The standard CE mark signifying a claim to conform to relevant directives

Because of the characteristic that the CE mark is to be placed on the product when the manufacturer claims conformity the New Approach Directives are sometimes known as 'CE Marking Directives'. When applying CE marking the manufacturer is indicating that all those directives which call for CE marking have been complied with. This places a considerable burden on the design team since they have to work out which directives will apply.

How do you know which directives apply to your project?

If you consult the list of directives shown in the table of Section 2.2.5 it will be clear that several of them might apply to a product depending on its make up.

Generally for a machine product the Machinery Directive will always apply but if the machine has an electrical power drive and a control system, the LVD and EMC Directives will also apply.

This workshop concentrates mainly on functional safety systems; those safety measures that are based on sensors and control systems that are design to ensure safe working of the machines.

2.2.5   List of New Approach Directives

How do you know which directive apply to your project?

If you consult the list of directives shown in the following table it will be clear that several of them might apply to a product depending on its make up.

Generally for a machine product the Machinery Directive will always apply but if the machine has an electrical power drive and an electronic control system the LVD and EMC Directives will also apply as indicated by Figure 2.4.

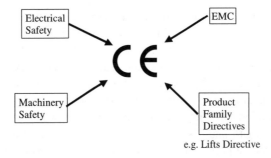

Figure 2.4
Typical scope of CE marking

This workshop concentrates mainly on functional safety systems; those safety measures that are based on sensors and control systems that are design to ensure safe working of the machines. These principles are found predominately within the scope of the Machinery Directive which has safety in use as its main objective.

The full list of directives and their related harmonized standards can be found on the Europa website http://www.newapproach.org.

396/EEC	Appliances burning gaseous fuels
00/9/EC	Cableway installations designed to carry persons
89/106/EEC	Construction products
89/336/EEC	Electromagnetic compatibility (EMC Directive)
94/9/EC	Equipment and protective systems in potentially explosive atmospheres (ATEX Directive)
93/15/EEC	Explosives for civil uses
95/16/EC	Lifts
73/23/EEC	Low voltage equipment (LV Directive)
90/385/EEC	Medical devices: active implantable
93/42/EEC	Medical devices: general
98/79/EC	Medical devices: in vitro diagnostic
90/384/EEC	Non-automatic weighing instruments
94/62/EC	Packaging and packaging waste
89/686/EEC	Personal protective equipment
97/23/EC	Pressure equipment
99/5/EC	Radio and telecommunications terminal equipment
94/25/EC	Recreational craft
98/37/EC	Safety of machinery (Machinery Directive)
88/378/EEC	Safety of toys
87/404/EEC	Simple pressure vessels

Deciding on the scope of the directives and how they are to be applied to a particular product can be a difficult task for a manufacturer and it is an area where specialist advice may be needed. Such advice is available from organizations that are accredited by national authorities to act as 'Notified Bodies' for the inspection and evaluation of products for compliance with EU Directives. For example, in the UK the British Standards Institution (BSI) offers such an advisory service (see www.bsi-global.com).

2.2.6   Requirements for CE marking and its implications

The products, which come under European Directives, and are to be placed on the market in the EU, must bear CE marking – it is a legal requirement. The affixing of CE marking to machinery by the manufacturer is to show its conformance to that all-relevant essential requirement as per 'New Approach' European Directives. CE marking is not a quality mark nor is it a mark of origin, and affixing it on machinery is only one of the several requirements that the manufacturer has to meet. The machine should also be accompanied by the relevant documentation showing its conformance to the provisions of Machinery Directive.

The CE mark shall be distinct, visible, legible and indelible. In case the manufacturer or the responsible person puts or does not put CE marking but fails to provide full information, if required to do so by enforcement authority, he leaves himself open to prosecution. *CE marking must not be affixed to safety components under the Machinery Directive.*

2.2.7 How can the technical requirements of a directive be defined?

The best, but not the only method, is to link the requirements to a recognized engineering standard. Hence the EU has adopted the concept of harmonized standards as the reference document wherever there is one available. For engineers working in machinery safety it is essential to have an outline understanding of the origin and structure of harmonized standards. Since there are approximately 700 of them and the numbers are still growing they present a vast body of knowledge and experience available to the designer.

2.2.8 Harmonized European standards

This term describes a large family of standards that have been accepted by all EU member states as representing the technical requirements of the EU Directives for health and safety aspects of manufactured products. These are drawn up by the two standards organizations CEN (Comité Européen de Normalisation) and CENELEC (Comité Européen de Normalisation Électrotechnique) as mandated by the EC Commission (see Figure 2.5).

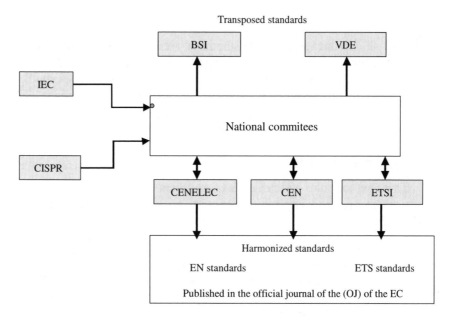

Figure 2.5
The process for creating harmonized standards

The standards are prefixed as EN Standards and are then transferred unchanged into the national standards of member states. By fulfilling such harmonized standards, there is an 'automatic presumption of conformity', i.e. the manufacturer can be trusted to have fulfilled all the safety aspects of the Directive as long as they are covered in the particular standard. This relationship was seen earlier in Figure 2.2.

2.2.9 The structure of harmonized standards

Figure 2.6 shows the harmonized standards framework used in the EU to provide support to compliance with the product directives.

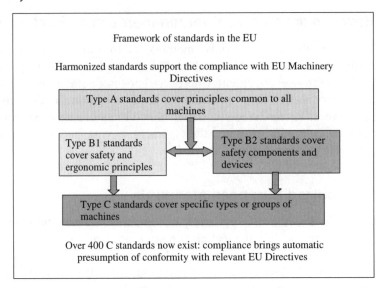

Figure 2.6
The EU standards framework

As shown in Figure 2.6 the standards are based on a set of general safety principles for all machines contained in the highest level or type A standards. The methods and devices commonly used to provide safety have been identified in the type B level standards. Type C standards are then compiled for particular type of machines incorporating principles or referencing methods defined in the higher levels.

Type A standards

These set the rules and principles for writers of more specific standards and for any design team to apply to any new machinery project. Two of the most widely known type A standards are:

1. *EN 292 Parts 1 and 2:*   Safety of Machinery. Basic terminology, general design principles. Part 1 mainly handles the risks to be evaluated and the design principles to be used to reduce the risks. Part 2 outlines the basic principles of machinery guarding, interlocking, E-stops, trip devices, safety distances, etc.
2. *EN 1050:*   Safety of Machinery, Principles of Risk Assessment. EN 1050 sets down methods for risk assessment that form the first essential stage in the development of protection systems for machinery.

Type B standards

Type B1 standards set down design requirements for safety techniques such as the provision of SRECs. Examples relevant to control engineers are: EN 60204: Safety of machinery – electrical equipment of machines Parts 1 and 2 and EN 954: Safety of machinery – safety-related parts of control systems – Parts 1 and 2.

Type B2 standards deal with widely used safety devices such as light curtain detectors and two hand controls. Examples are: EN 418 for E-stop switches and EN 61496 for the application of light barriers.

Type C standards

A large number of type C standards have been produced to deal the hazards of specifically identified types of machines. The most common of these are the manufacturing plant machines beginning with power presses. Because these machine type standards have been prepared using the foundation of type A and B standards they will generally have a consistent basis for the safety requirements defined in their texts. For example, they will base any devices suggested for safety guarding or E-stops to the relevant type B standard.

Obviously any new technology-driven developments influencing the type A or B level standards will have to be carefully introduced to this hierarchical system. Changing the ground rules is not going to be easy.

If there is no harmonized standard there is an order of preference for other standards which runs as follows:

- *Harmonization documents (HDs for short):* CEN and CENELEC draft harmonization documents if transposition into identical national standards is unnecessary or not feasible. This happens when there are certain national divergences. In practice CEN has not been drafting harmonization documents for several years. CENELEC has published many, and tends to transform them into European standards when they come up for revision.
- *European standards:* These are standards established by the European standards bodies to meet industrial or commercial requirements but with no particular link with a 'New Approach' Directive or legal constraint. Any national standard dealing with the same subject must be withdrawn and replaced by the transcription of the European standard. European standards, whether or not harmonized, are available only through the member states' national collections. They are identified by the letters 'EN' in the name of the standard. For example: BS EN 954-1 is the version of EN 954-1 published by the British Standards Institution.
- *National or international standards:* These can be of national origin, European origin (see the previous categories) or international origin. Very often national standards will also take over the contents of international standards. This is frequently the case with the wide range of IEC standards and, for example, the functional safety standard IEC 61508 is published in Australia as AS. 61508 and in the UK as BS IEC 61508.

In the absence of any recognized standard for a particular application the safety requirements of the Directive can still be met by employing the basic principles for design and safety as laid down in the type A standards. Effectively you can do the job from 1st principles. This should still be acceptable for demonstrating compliance, provided the reasoning process is properly recorded and validated.

2.3　The European Machinery Directive

2.3.1　Introduction

The Machinery Safety Directive impacts all machinery that is to be made or supplied in the EU even if it is imported from outside the EU. For imported machines the Directive also applies to re-furbished used machinery. Hence any machinery designer/builder wishing to supply to an EC country must be able show conformance to the EC Machinery Directive.

For machinery already existing in the EC that has escaped the Machinery Directive the end user remains responsible for its safety features and its end use through the requirements of the Use of Work Equipment Directive.

The Machinery Directive 98/37/EEC aims to achieve an open market for machinery supply in Europe. Its effect in respect of safety should be that codes of practice would be the same throughout Europe. It also has the effect of requiring that all new and used machinery imported from outside the EU shall comply with the same practices.

2.3.2   Finding the Directive

The official reference description for the Machinery Directive is: Directive 98/37/EC of the European Parliament and of the Council of 22 June 1998 on the approximation of the laws of the Member States relating to machinery Official Journal L 207, 23/07/1998 pp. 1–46. You can obtain a copy of the Directive by downloading from the Europa website www.newapproach.org/directive.asp.

2.3.3   To what does the Directive apply?

The scope of the directive makes it very clear that most forms of machinery are covered: Machinery is defined as one of the following:

An assembly of linked parts or components, at least one of which moves, including, with the appropriate actuators, control and power circuits, joined together for a specific application, in particular for the processing, treatment, moving, or packaging of a material.
e.g. A pump, motor, and starter unit are not machines as individual components, but they are integral components of an independently functioning machine capable of moving a material (fluid).

An assembly of machines which, in order to achieve the same end, are arranged and controlled so that they function as an integral whole.
e.g. A pumping skid containing two or more of the assemblies above tied into a common outlet line for the purpose of boosting flow volume is also a machine.

Interchangeable equipment modifying the function of a machine which is supplied for the purpose of being assembled with a machine (or a series of different machines or with a tractor) by the operator himself in so far as this equipment is not a spare part or a tool.
e.g. Farm equipment, which modifies the function of a tractor when attached.

The Machinery Directive also covers safety components for machinery, defined as:

Components which are supplied separately to fulfill a safety function when in use and the failure of malfunctioning of which endangers the safety or health of exposed persons.
e.g. A limit switch for a safety guard or a light barrier at the opening of a press.

Thus, the term machinery covers any equipment, whether for domestic, commercial or industrial applications, that has parts actuated by a power source other than manual effort. It covers from a basic machine up to a complete plant.

However, there are some exceptions; the directive includes list of machine types that are excluded from its scope for various reasons. These types may be covered by other directives or by existing specialist laws. The list is shown in Appendix A to this manual.

For example: The Low Voltage Directive (73/23/EEC) will apply where the risks are mainly of an electrical origin and there may be no need to apply the Machinery Directive in such cases.

2.3.4   Contents and structure of the Machinery Directive

The following table gives a summary of the 14 Articles and 7 Annexes of the Machinery Directive:

Articles	Areas Covered	
1–7	Application area, selling, marketing, freedom of movement, essential health and safety requirements	
8–9	Conformity assessment procedures	
10–12	Protection against arbitrary fulfillment	
13–14	Coming into force, transitional regulations, cancellation of the regulations	
Annex	**Areas Covered**	**Articles**
I	Essential health and safety requirements relating to the design and construction of machinery and	3
	Interchangeable equipment	5
	Safety components	10
II	Contents of	
	1. EU Declaration of Conformity for machinery and	4
	Interchangeable equipment	5
	Safety components	8
	2. Manufacturers declaration for	
	Specific components of the machinery	4
	Non-functioning machines	
III	CE marking	10
IV	Types of machinery and safety components where the procedure refers to Article 8 must be applied	
V	EC Declaration of Conformity for machinery and Interchangeable equipment Safety components	8
VI	EC type examination for machinery and Interchangeable equipment Safety components	8
VII	Minimum criteria to be taken into account by member states for the notification of bodies	9

It is easy to navigate the above structure of Articles and Annexes. The Directive contains a great deal of straightforward information about the principles of machinery safety as well as details on all the steps that must be taken to achieve conformity with the directive.

In this workshop we have to limit our attention to the technical aspects of the directive, so Annex I and Annex IV are of special interest. We shall also outline the conformity declaration in Annex V because this has a big impact on how any machine building project is conducted.

2.3.5   Annex I and the EHSRs

Article 3 declares that all machines are to be subject to requirements set out in Annex I. So we need to take a look at Annex I. Figure 2.7 describes how the annex is divided into groups and sections, and outlines the contents of each section.

The important thing to note here is that this structure and its contents are very closely followed by the national regulations in the UK defined in the Supply of Machinery Act and these are the same requirements that are to be found in the national laws of other European Union members.

From Figure 2.7 you can see that any particular aspect of machine safety can be found quickly in one of the general sections.

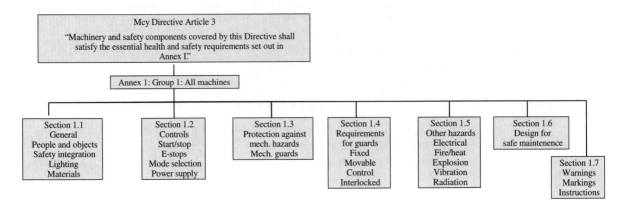

Figure 2.7
Annex I, Sections 1.1 – 1.7

Further parts of Annex 1 list the EHSRs for specialized machines or special aspects of machines. These groups are:

- *Group 2:* Agri-foodstuffs machinery, portable and/or hand-guided machines and woodworking or similar material-working machines
- *Group 3:* ESHRs to offset particular hazards due to mobility of machinery
- *Group 4:* ESHRs to offset particular hazards of lifting operations
- *Group 5:* ESHRs for machinery intended for underground work
- *Group 6:* ESHRs to offset particular hazards due to the lifting or moving of persons.

Requirements under these groupings cover well-known hazards and describe known techniques for risk reduction. For example, in Group 2, Paragraph 2.3, Section C states: 'the machinery must be equipped with an automatic brake that stops the tool in a sufficiently short time if there is a risk of contact with the tool whilst it runs down'.

2.3.6   Linking Annex I EHSRs to standards

The next step will be to examine the detailed requirements in any one of the boxes or sections we have illustrated. This is where the harmonized standards come into the picture. If you read any one of the above sections of Annex I it will give you a general view of what is required but there is a need for more guidance in detail. In this case, standard EN 292 covers many of the subject items listed in the above sections. In fact, we can map the parts of EN 292 in Figures 2.8 and 2.9.

The linkage between EN 292 and the EHSRs is further strengthened by the fact that Annex I of EN 292-1 is complete copy of the EHSRs laid down in Annex I of the Machinery Directive. Therefore for practical purposes of safety design you can use EN 292 as the information source for all general EHSRs.

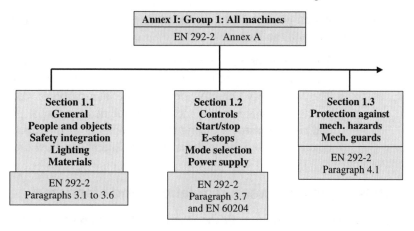

Figure 2.8
Annex I, Sections 1.1 – 1.3

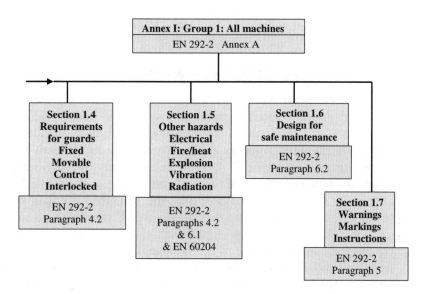

Figure 2.9
Annex I, Sections 1.4 – 1.7

So with the help of EN 292 and several other harmonized standards (notably EN 60204 for electrical safety), the designer of a machine can systematically work toward compliance with the Directive. In summary, the EHSRs in Annex I of the Directive require the manufacturer to ensure safety of the machine by applying the following principles, in the order given:

- Eliminate or reduce risks as far as possible (inherently safe machinery design and construction)
- Take the necessary protection measures in relation to risks that cannot be eliminated
- Inform users of the residual risks due to any shortcomings of the protection measures adopted; indicate whether any particular training is required.

The protection goals must be responsibly implemented in order to fulfill the demand for conformance with the Directive.

- The manufacturer of a machine must prove that the basic requirements are fulfilled. This proof is made easier by applying harmonized standards.
- The Machinery Directive demands the integration of safety as early as the design process. In practice this means that the designer must perform a hazard analysis and risk assessment during the development of the machine so that the measures developed from the analysis and assessment can flow directly into the design.

2.3.7   Relationship between regulations and standards

Do we have to comply with the standards? It is important to understand that standards have no legal status unless they are referenced by legislation or legal decree. In the 'new approach' Directives, technical standards are always applied voluntarily and failure to comply with them is never in itself a fault. However, they do represent the state of the art and very often legislation will expect compliance with the best current practices. So if you don't comply with what appears to be a relevant standard you have to justify the position on merit, and this can be a lot harder.

2.3.8   The role of type C product standards

The standards organizations CEN and CENELEC have carried out the major program of producing machine-specific standards based on the principles laid down in type A and B standards. As a result there are now over 750 type C standards available for use by manufacturers in designing for conformity with the EU Directives.

Conformity with a type C standard that correctly covers your particular machine will bring an automatic presumption of compliance with the EHSRs of the Machinery Directive. You can then assume that the machine complies with the type A and B standards that have been used to guide the type C standard.

Clearly the first step in compliance for a manufacturer is to find out if there is a type C standard applicable to the product. The listing of type C standards under-goes frequent updating and the source for the master list is the Official Journal (OJ) of the EC. See www.NewApproach.org/directiveList.asp. The Europa website http://europa.eu.int/comm/enterprise/index_en.htm. provides guidance and a convenient summary list of all standards relevant to machinery.

2.3.9   Annex IV and the special categories

The second area of the Directive we need to consider is the position of Annex IV (also known as Schedule 4) machines. Here the Directive sets up a list of machine types that are known for particular types of hazards that are severe in nature. The Directive requires that these types are to be subject to third-party assessment of the safety of the design and the protection measures that have been provided by the designer.

The same requirement applies for certain critical types of safety components also listed in Annex IV. The heading to Annex IV reads 'Types of Machinery and Safety Components for which the Procedure Referred to in Articles 8(2)(b) and (c) must be Applied'.

Then follows a listing of machine types that require the additional approval stages that are required by Article 8. One of the difficulties facing the machinery and safety system builders is that new additions to the Annex IV list are made from time to time

Hence the rules regarding conformity will change as a machine type becomes listed. It is essential to check the listing of machine types and related standards by reference to www.NewApproach.org/directiveList.asp.

Please see Appendix A in this manual for details of Article 8 and a typical listing from Annex IV. Here next is a brief outline of the procedure for the assessment of conformity.

2.4 Conformity procedures

All procedures for achieving conformity are set out in Annex VI of the Machinery Directive Machines are identified as either of a non-specific type (having no directly identified hazardous features) or are specially identified as hazardous in the Annex IV list as noted above. All machines that are not listed in Annex IV are subject to the standard procedure.

2.4.1 Conformity for non-listed machines

Machines not listed in Annex IV can be declared to be in conformity by the manufacturer based on evidence compiled in a technical file (see below). There is no obligation to involve a third party such as testing specialist testing authority (Notified Body) but the technical file must be kept available by the manufacturer for examination by the end user.

The machine has to conform to either Product Standards Type C or conform to ESHRs regulations in which case the A and B type standards can be used to satisfy the requirements. Figure 2.10 shows the option of either fixing the CE mark or providing a 'Declaration of Incorporation'.

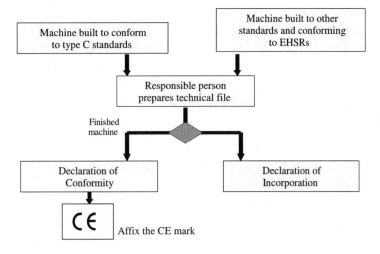

Figure 2.10
CE routes to conformity for non-listed machines

2.4.2 Declaration of conformity

When a machine is to be sold as a standalone item ready for use the manufacturer must issue a Declaration of Conformity. This is issued after the machine is finished and ready for installation and use. A Declaration of Conformity must:

- State the name and address of the manufacturer or (in the case of machinery made outside the EEA) of the importer
- Contain a description of the machinery, its make, type and serial number
- Specify which standards have been used in design and the manufacture (if any)

- Indicate all relevant European Laws (Directives) with which the machinery complies
- State details of any notified body that has been involved
- Be signed by a person with authority to do so.

The Declaration of Conformity must be in the recognized language of the country where it is to be used.

2.4.3   Declaration of incorporation

If a machine is designed to be incorporated into other machinery, it might not have CE marking fixed to it. It should be manufactured to be as safe as possible and be provided with a Declaration of Incorporation. Instructions on safe installation and use should also be provided. When the machine is fitted into the assembly line, particular attention must be given to any hazards which may have been caused by the machine being fitted into the line. For instance, additional guarding or other controls may be required.

Once the machine has been fitted and the whole line is safe, the technical file should be completed and either the machine or the whole line should have CE marking. This can be done by a project manager (e.g. the installer, assembler or the manufacturer) but in many cases the end user can do it, particularly if it is a small company.

The declaration should contain similar information to that contained in the Declaration of Conformity, but importantly, it must state that the machinery should not be used until: *the machinery into which has been incorporated or the assembly to which it has been added has been declared to conform fully with the legal requirements.* CE marking should only take place at the end of incorporation or assembly.

Clearly, if you obtain a machine as a subsystem to build up a complex machine it is important to have the Declaration of Incorporation from the supplier as this will simplify your task of creating the technical file and declaring conformity.

2.4.4   Contents of the technical file

Manufacturers are required to draw up a 'technical construction file' for the machinery they make. Technical files are to demonstrate how machinery meets relevant essential health and safety requirements and, as such, are useful for manufacturers and for the national enforcing authorities.

Annex V of the Directive describes the task of the manufacturer as follows:

Before drawing up the EU Declaration of Conformity, the manufacturer, or his authorized representative in the community, shall have ensured and be able to guarantee that the documentation listed below is and will remain available on his premises for any inspection purposes:

(a) a technical construction file comprising:
  - *an overall drawing of the machinery together with drawings of the control circuits*
  - *full detailed drawings, accompanied by any calculation notes, test results, etc., required to check the conformity of the machinery with the essential health and safety requirements*
  - *a list of the essential requirements of this Directive, standards, and other technical specifications which were used when the machinery was designed*
  - *a description of methods adopted to eliminate hazards presented by the machinery*

> – *if he so desires, any technical report or certificate obtained from a competent body or laboratory (1)*
> – *if he declares conformity with a harmonized standard which provides therefore, any technical report giving the results of tests carried out at his choice either by himself or by a competent body or laboratory (2)*
> – *a copy of the instructions for the machinery.*
> (b) *for series manufacture, the internal measures that will be implemented to ensure that the machinery remains in conformity with the provisions of the Directive.*

The manufacturer must carry out necessary research or tests on components, fittings or the completed machine to determine whether by its design or construction, the machine is capable of being erected and put into service safely.

It is instructive to look at how the UK HSE describes the duties of machinery buyers when obtaining a machine from a supplier. Readers are recommended to obtain a copy of the publication no. INDG271 04/98 C200 'Buying New Machinery' by download from the HSE website: http//www.open.gov.uk/hse/hsehome.htm.

2.4.5   Conformity for Annex IV listed machines

For the hazardous machinery or safety component as listed in Annex IV, conformity requires the involvement of an authorized body such as an accredited testing and inspection company, known as a 'Notified body'. The degree of involvement of the Notified body depends on the extent to which type C standards exist for the application and on the decision of the manufacturer on the extent of external approval it is seeking. This last point will probably depend on the degree of confidence the manufacturer may have in his knowledge of the conformity requirements. Figure 2.11 maps the procedures.

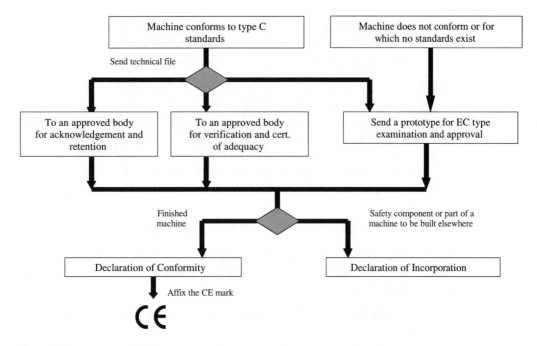

Figure 2.11
CE conformity routes for Annex IV listed machines

If harmonized standards exist for the working machinery or safety components that cover the entire range of requirements, two cases are possible:

1. The manufacturer sends the technical documents (as listed in Annex VI of the directive) to a Notified body; this body confirms receipt and archives the documents.
2. The manufacturer requests the Notified body to check his technical documents in respect of compliance with the harmonized standards. If this is the case, the body provides the manufacturer with a certificate about the compliance with these standards.

Clearly the choice between item (1) and item (2) depends on the degree of confidence that the manufacturer has in his application of the design procedures and standards.

2.4.6   EC type examination

Where there are no harmonized standards for machinery, or, if the machine or parts are not built to the standards, the manufacturer must make the machine and the technical documents available to the Notified body so that an EC type examination can be performed.

In all cases the manufacturer makes out the Declaration of Conformity for the related product on its own liability, and thus undertakes the obligation that the equipment and protective devices have been designed in compliance with the standards.

2.4.7   Safety components in Annex IV

Also included in Annex IV are safety components, i.e. devices that provide a safety function. These are of particular interest to our workshop where we focus on the instrument and electrical practices needed to provide functional safety in the machines. The safety devices include but are not limited to:

- E-stops and movable guards
- Electrosensitive devices to detect persons to ensure their safety (non-material barriers, sensor mats, electromagnetic detectors, etc.)
- Logic units which ensure the safety functions of bimanual controls.

There are some important implications here. If you are going to provide a safety-related control system and incorporate sensing devices such as light curtains or safety mats you will need to ensure that these devices are compliant with the Machinery Directive. Either they should have CE marking themselves or have a declaration of incorporation and the route to this position includes the approval by an accredited testing body as we saw above. Hence safety components will normally be sold with CE marks and testing authority badges such as BSI or TUV.

2.4.8   UK implementation of Machinery Directive

As a good example of how the Directives are implemented at national levels in the EU we have seen earlier that the UK has two major laws in force relevant to the supply and use of machinery. These are:

1. The Supply of Machinery (Safety) Regulations 1992 (known as 'the supply law').
2. The Provision and Use of Work Equipment Regulations 1998 (PUWER) (known as the user law). This regulation replaces the earlier 1992 version.

The supply law requires manufacturers and suppliers to ensure that machinery as supplied to the user is

- Safe by design as far as practicable
- Safe to use through the integrity of its safeguarding measures
- In compliance with applicable directives and has CE marking to show this.

The user law requires employers to:

- Provide the right kind of safe equipment to use at work
- Ensure that it can be used correctly
- Keep it maintained in a safe condition.

Enforcement of these laws in the UK is the responsibility of the UK Health and Safety Executive (HSE). The HSE is a most valuable source of practical advice for suppliers and users of machinery in respect of safety requirements. See Appendix 2 for references to important publications available from HSE. The regulations are enforceable for all machinery supplied from January 1, 1995, in the United Kingdom, but not to machinery supplied before this date.

2.4.9   UK safety regulations for supply of machinery

Scope of the regulations

The Supply of Machinery (Safety) Regulation: SI 3073/1992 came into force on 1st January 1993, with a 2-year transition period to enable industry to meet the requirements. Thus from 1st January 1995 most of the machinery has to satisfy wide-ranging ESHRs for design and construction of machines.

Responsibilities of the supplier

Here is a summary of the supplier's responsibilities based on guidelines issued in the HSE guide 'Supplying New Machinery'. The supplier must make sure that the machinery is safe for use as per Machinery Directive before he supplies it in anywhere in EEA. To do this they should first ensure that:

- Machinery meets relevant essential health and safety requirements (these are listed in detail in the regulations) which include the provision of sufficient instructions.
- A technical file for the machinery has been drawn up, and in certain cases the machinery has been type-examined by a notified body.
- There is a Declaration of Conformity (or in some cases a Declaration of Incorporation) for the machinery, which should be issued with it.
- There is CE marking affixed to the machinery (unless it comes with a Declaration of Incorporation).

There are several practical checks that suppliers can make – these are given in the checklist as below:

- Look for obvious defects, such as missing guards or other safety devices, or inadequately protected electric wiring.
- Check that known risks (including risks from fumes or dust, noise or vibration) from the machinery when it is in use will be properly controlled or that there is information on how they can be controlled.

- Make sure that the manufacturer has provided instructions for safe installation, use, adjustment and maintenance, and that these are in English if the machinery is for use in the UK (some maintenance instructions may be in another language if staff from the manufacturer are to carry out specialized maintenance).
- Check that data about noise and vibration emissions have been provided were appropriate.
- Make sure that any warning signs are visible and easy to understand.
- Check that you have a properly completed Declaration of Conformity, or a Declaration of Incorporation if appropriate.
- Check for CE marking.
- Check with the manufacturer if you have concerns about any of the above matters.

If you are a importing machinery from outside the EEA, you will need to make sure that the machinery meets the requirements of the law in full. This machinery must still meet the Supply of Machinery (Safety) Regulations 1992 and EU Machine Directives. Since you are importing it directly from outside the EEA, you take on the legal responsibilities of the manufacturer. So you need to make sure either that the manufacturer has met the relevant requirements or that the supplier meets them.

2.5   Other 'New Approach Directives'

There are two other related directives that are generally relevant to the supply of machinery.

2.5.1   The Low Voltage Directive

The Low Voltage Directive (73/23/EEC of 1973 and modified in 1994) (LVD) states that all electrical equipment must be safe in usage except for residual risks. The LVD applies to equipment with a voltage rating of between 50 and 1000 V AC and between 75 and 1500 V DC

The electrical equipment should be safe from hazards of insulation failure and shock being the prime requirement, which can be achieved through proper insulation, testing and earthing. As in the case of the Machinery Directive compliance to standards is the best route. Essentially it should meet the following safety objectives:

- General conditions for safe usage
- Protection against hazards arising from equipment
- Protection against external influences.

The requirement of technical file and conformity are the same as for Machinery Directive and enforcement is to be ensured by local authorities.

Some of the key standards written to support LVD directives are:

- *EN 60204-1:*   Safety of Machinery – electrical equipment of machines
- *EN 60335-1, Part 1:*   General requirements – safety of household appliances and similar goods (There are approximately 100 parts to this standard to cover the range of appliances.)
- *EN 60947-1, Part 1:*   General requirements – low-voltage switchgear and control gear
- *EN 60598-1, Part 1:*   General requirements – luminaries
- *EN 60730-1, Part 1:*   General requirements – automatic electrical controls for household use, and similar.

The LVD extends to cover any electrical device or appliance and is very far reaching. As noted earlier it is essential to consult the list of harmonized standards related to this Directive when planning a project with electrical equipment destined for the European market.

As an example of the complexities of the LVD consider the case of an electronic product such as a Notebook PC. The unit operates at less than 50 V internally but it is provided with a power supply module, internal or external that operates at mains voltage of typically 230 V AC. Hence the power supply unit must meet LVD requirements in order to ensure personal safety and to ensure that the low-voltage portions of the PC cannot become a hazard through defects in the power unit.

2.5.2   Which Directive? MD or LVD

Which directive should apply to electrical equipment used in machines? Article 1.5 of the MD states that where the risks are mainly of an electrical origin the machinery shall be covered exclusively by the LVD. EU guidance suggests that in case of doubt the manufacturer should apply EN 1050 on Risk Assessment to determine whether the risk arises from electric shock.

Machinery intended for the consumer and for some light commercial uses can be dealt with under the LVD because this Directive includes for mechanical risks and is supported by standards that include for these.

So the manufacturer decides this issue with the aid of a risk assessment where necessary. Whatever the outcome, the CE declaration must state which directive has been considered for compliance.

For the machinery safety applications we shall be particularly concerned that the electrical equipment, drives, actuators, measuring instruments and the electrical control panels are all built to appropriate type C standards. In practice most of our needs are covered by EN 60204-1: 1997. Since this is the same standard that is harmonized for compliance with the Machinery Directive it should not present too much of a problem to designers.

2.5.3   The EMC Directive

This Directive seeks to ensure that manufactured equipment does not emit unacceptable levels of EM radiation that will interfere with other electrical and telecommunication devices. It also ensures that products have a necessary level of immunity to normal levels of EM radiation.

The key standards written to support EMC directives are:

- *EN 50081-1, Part 1:* Electromagnetic compatibility – generic emission standard (residential, commercial and light industry)
- *EN 50081-2, Part 2:* Electromagnetic compatibility – generic emission standard (industrial environment)
- *EN 50082-1, Part 1:* Electromagnetic compatibility – generic immunity standard (residential, commercial and light industry)
- *EN 50082-2, Part 2:* Electromagnetic compatibility – generic immunity standard (industrial environment)
- IEC 1000 Good EMC practice for installers.

For safety-related control systems the immunity aspect is an important issue since the integrity of the safety function could be challenged by electrical interference from the machine power drives and from the industrial environment. At a seminar on machinery safety held in the UK in 2002 the specialist company Compliance Engineering offered the following comments on the effects of EM radiation on safety.

Limitations of EMC Directive

- There is a common misperception that all that is needed to control EM interference for all purposes in the EU is to use apparatus which is CE marked and declared compliant with the EMC Directive.
- The EMC Directive only covers normal operation and does not cover reasonably foreseeable faults, environmental extremes, operator errors, maintenance situations, or misuse – all considerations which are essential for functional safety.

EMC Directive vs Machinery Directive

- Almost all the EMC standards harmonized under the EMC Directive either explicitly or implicitly exclude safety considerations.
- All the EMC Directive harmonized standards cover a restricted number of EM disturbances, and their limits allow a finite probability of incompatibilities.
- EMC testing can never prove that an apparatus is totally safe because EM disturbances have a statistical distribution and it is impossible to know what the most extreme level of a disturbance of a given type may be.

(From a presentation by John McAuley, Compliance Engineering Ireland Ltd. For details see: IEE Paper on EMC and Functional safety: www.iee.org.uk/PAB/EMC/core.htm or contact: www.cei.ie.)

The above comments lead to the conclusion that a risk assessment carried out on machinery with built-in safety systems should take into account the possibilities of external radiation sources such as arcing contactors or welding machines.

2.5.4   Supply side directives summary

The Machinery Directive calls for a complete risk assessment for any type of machinery.

- Harmonized standards provide access to design methods based on past experience and that are widely accepted in many parts of the world.
- Conformity procedures encourage a responsible approach to providing safety in any machine.
- Other directives must be considered when seeking to achieve conformity to EU product directives.
- CE marking indicates that the manufacturer or supplier claims conformity to relevant directives.
- The manufacturer/supplier must decide and record which directives apply.

2.6   User side directives: workplace health and safety legislation

As we have noted in Chapter 1 most countries set up a framework of legislation for health and safety in the workplace. As the EU developed, it set up a similar structure in the form of the Article 118 Directives, which are known as the Framework Directives. These are shown in Figure 2.12.

Included in these is the 'The Use of Work Equipment Directive' which is the one most directly relevant to machinery safety. The aim is to improve the level of safety of working equipment. It requires employers and operators of equipment to comply with essential health and safety requirements very much as defined for the supply side.

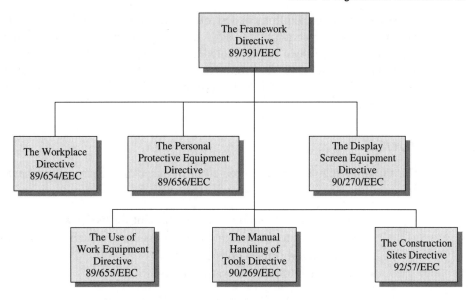

Figure 2.12
EU Framework Directives for health and safety

2.6.1   UK implementations of the Framework Directive

The Framework Directive is implemented in the UK as set of regulations known as 'The six pack' as shown in Figure 2.13. These provide a typical arrangement of protection measures for workers and similar sets of laws can be seen in many industrialized countries. In particular it will be instructive to look at the regulations affecting the use of machinery.

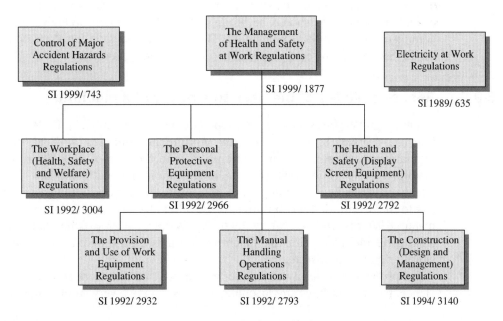

Figure 2.13
Workplace Laws in the UK implementing the Framework Directive

The Use of Work Equipment Directive was amended in 1995 to include mobile work equipment, lifting equipment and the inspection of work equipment. All these requirements have been implemented in the UK either through the general Health and Safety at Work Act (HSWA), the Management of Health and Safety at Work Regulations 1992 (MHSWR) or through the Provision and Use of Work Equipment Regulations 1998 (PUWER).

2.6.2 PUWER

The PUWER regulations are directed at the user of equipment but have implications for the suppliers. They require employers to:

- Provide the right kind of safe equipment for use at work
- Have regard to the working conditions under which it is to be used
- Ensure that it is used correctly and
- Keep it maintained in a safe condition.

PUWER has some 20 odd subsections dealing with the safety aspects of machinery use and defining the responsibilities of the 'employer' and the 'user'. We have placed a brief summary of these requirements in Appendix C because they are informative and will serve as a useful guide for anyone planning a machinery application project. UK readers are recommended to obtain a copy of the guidebook 'Safe Use of Work Equipment' published by the UK Health and Safety Commission and available through HSE books. This book integrates the PUWER regulations with guidance to form what is known as an 'Approved Code of Practice' or AOCP.

The AOCP appears to perform a similar role to the type A harmonized standards referenced by the Machinery Directive. The AOCP provides good practical guidelines for each regulation defined in PUWER. The guidance states, 'If you follow the advice you will be doing enough to comply with the law in respect of those specific matters in which the Code gives advice.' In general terms, the regulations require that the equipment be safe enough and employees should not be exposed to hazards at workplace. The risks if cannot be eliminated should be brought down to minimum tolerable level and suitable administrative controls implemented. These regulations give legislative power to enforce the EHSRs make prosecution possible on non-implementation.

2.6.3 Responsibilities of the user as a buyer of a machine

If an employer buys new equipment (including machinery) PUWER also requires him, as a user, to check that the equipment complies with all the supply law regulations that are relevant. The employer in the role of user has to ensure that the manufacturer's claim that the machinery complies with the law is properly and adequately checked for compliance for safety during operation, before it is used. He has to make sure that the supplier (or installer) has given him information on how the machine works and its safety features. With smaller off-the-shelf machinery, this should be included with the machine. With complex or custom-built machines the supplier may demonstrate this.

The prime responsibility of the user is to make sure that work equipment is safe to use. Hence, before user buys any machinery he has to give a very careful thought to:

- Where and how the machine will be used?
- What it will be used for?

- Who will use it (skilled employees, trainees)?
- What risks to health and safety might result?
- Comparing how well health and safety risks are controlled by different manufacturers.

The answers to above can help user to decide which machine may be most suitable, particularly if a standard machine off-the-shelf is being bought. In case a more complex or custom-built machine is being bought, the requirements should be discussed with potential suppliers. They can often advise on the options available.

For a custom-built machine, the opportunity to work with the supplier is to be utilized to design out the causes of injury and ill health. Time spent now on agreeing the necessary safeguards, to control health and safety risks, could save substantial time and money later.

The following could serve as a checklist for the user to follow at the time of purchasing and on delivery of a machine:

- Check that the equipment being purchased has CE marking (where necessary).
- Check for the copy of EU Declaration of Conformity.
- Check that the supplier has explained what the machinery is designed to be used for and what it cannot be used for (unless this is off-the-shelf machinery).
- Check to see if you as a user think that it is safe.
- Make sure a manual has been supplied which includes instructions for safe use, assembly, installation, commissioning, safe handling, adjustment and maintenance.
- Make sure the instruction manual is written in the local language of use. (The maintenance instructions may however be written in another language if specialized staff from the manufacturer or supplier will carry out maintenance.)
- Make sure information has been provided about any remaining/residual risks from the machine, and the precautions you need to take to deal with them. These may include electrical, hydraulic, pneumatic, stored energy, thermal, radiation or health hazards.
- Check that data about noise and vibration levels have been provided and, where necessary, explained.
- Ensure that all warning signs are visible and easy to understand.
- For a complex or custom-built machine, arrange for a trial run so that all the safety features and how they work can be shown and understood.
- Check to see for overall safety of the machine.
- Make sure that any early concerns about the safety of the machine are reported to the supplier.

2.7   Some machinery safety standards

This module concludes by looking again briefly at some of the most important standards used in the general practice of machinery safety. These are the standards we are going to be seeing more of as we go through the stages of developing the essential safety functions required to protect against machine hazards.

2.7.1 Type A standards – basic standards

These provide essential information for all machine builders. Generally there are three standards, which relate to machine safety:

1. *EN 414*. Safety of machinery: Rules for the drafting and presentation of safety standards. This defines the way standards are to be written.
2. *EN 292, Parts 1 and 2*. Safety of machinery: Basic concepts, general principles for design. This defines the concepts of machine safety and specifies the general principles and techniques to help machine designers achieve safety. It incorporates in Annex I in the EHSRs defined by the Machinery Directive.
3. *EN 1050*. Safety of machinery: Principles for risk assessment. This defines how to assess the risk of injury or damage to health so that appropriate safety measures can be selected.

2.7.2 Type B standards – group standards

B standards are subdivided into two groups:

1. *Group B1:* These cover higher-level safety aspects for design and are always applicable. For example, ergonomic design principles, safety distances from potential sources of danger, minimum clearances to prevent crushing of body parts. Examples of these are EN 294 on safety distances and EN 563 on temperatures of touchable surfaces.
2. *Group B2:* These cover safety components and devices for various machine types. These are applied when required. For example, E-stop equipment, two-hand controls, interlocking/latching, non-contact protective devices, safety-related parts of controls. EN 281, on the design of pedals, is an example.

Type B standards are as listed below:

European Standard's Number	International Standard's Number	Title of Standard
EN 61496-1	IEC 61496-1	Safety of machinery – Electrosensitive protective equipment – Part 1: General requirements and tests
PrEN 61496-2	IEC 61496-2	Part 2: Particular requirements for equipment using active optoelectronic protective devices
EN 61496-3	IEC 61496-3	Part 3: Particular requirements for equipment using active optoelectronic devices responsive to diffuse reflection (AOPDDRs)
EN 999	ISO 13855	Safety of machinery – The positioning of protective equipment in respect of approach speeds of parts of the human body
EN 294	ISO 13852	Safety of machinery: safety distances to prevent danger zones from being reached by the upper limbs

European Standard's Number	International Standard's Number	Title of Standard
EN 954-1	ISO 13849-1	Safety-related parts of control systems – Part 1: General principles for design
EN 954-2	ISO 13849-2	Part 2: Validation
EN 60204-1	IEC 60204-1	Electrical equipment of machines – Part 1: General requirements
EN 1088	ISO 14119	Interlocking devices associated with guards – Principles for design and selection
EN 953		Safety of machinery: Design and construction of guards
EN 574	ISO 13851	Two-hand control devices – Functional aspects: principles for design
EN 418	ISO 13850	E-stop equipment, functional aspects: principles for design
EN 1037	ISO 14118	Prevention of unexpected start-up. Safety of machinery. Isolation and energy dissipation
EN 811		Safety of machinery. Safety distances to prevent danger zones being reached by the lower limbs.

Standard EN 954-1 is of particular interest here to control and electrical engineers because it provides a method of classifying or grading the safety-related parts of control systems used in machinery applications. It classifies safety control functions into performance categories numbered 1, 2, 3, or 4 depending on the severity of their risk reduction tasks. It has close similarities to the safety categories used in safety instrument systems used for process control where they are known as SILs. We shall take a closer look at this standard and its principles in Module 4 of the workshop.

2.7.3 Type C standards – product standards

These identify specific types or group of machines and involve the machinery-specific standards. These inform machine manufacturers and users about the specific safety precautions they should take and safety devices they should use, e.g. for machine tools, wood-working machines, elevators, packaging machines, printing machines, etc.

A sample list of type C standards is shown here:

European Standard's Number	International Standard's Number	Title of Standard
EN 692		Mechanical presses; safety
EN 693		Hydraulic presses; safety
EN 12622		Hydraulic press brakes; safety
EN 775	ISO 10218	Manipulation industrial robots; safety
EN 1010	ISO 1010	Technical safety requirements for the design and construction of printing and paper converting machines
EN 11111	I SO 11111	Safety requirements for textile machinery

2.8 Regulations and standards in the USA

Here we take a brief look at the essential points of the US practices for machinery safety. Generally there are no radical differences in technical practices between the USA and Europe but we need to be aware of the equivalent structures and codes. When it comes to interchange of machinery via import or export it is essential that those involved know about the relevant codes of practice.

For example:

- US machine builders must comply with US regulations for the home market but must also meet EN or IEC standards when exporting into Europe.
- Companies producing machines for the US market must be familiar with US regulations for safety of machines and for certification of safety components.

In this workshop we cannot cover the potential differences that may arise between USA and EU in electrical or mechanical construction practices. We will limit ourselves to the basic safety issues and the features of SRECs.

2.8.1 US regulatory bodies

In the United States, the Occupational Health and Safety Association (OHSA) is a US Government body responsible for defining and enforcing requirements for maintaining a safe workplace environment.

- OHSA regulates standards for safety of persons
- OHSA is responsible for defining and enforcing requirements for a safe workplace environment.

Before OHSA (1970), industry used voluntary consensus based on standards produced through organizations such as:

- American National Standards Institute (ANSI)
- Underwriters Laboratories (UL)
- National Fire Protection Association (NFPA).

OHSA regulations now incorporate in law many of the safety practices laid down in the industry consensus standards. This is known as IBR, incorporation by reference.

2.8.2 Incorporation of safety standards into US law

The first standard for safe use of machinery was ANSI B11.1 – 1926. It has provided the foundation for many subsequent US machinery safety standards. The ANSI B11 series now carries many machine-specific safety standards, effectively performing the same role as EU type C standards.

- ANSI B11 standards cover applications at the machine level
- UL produces standards for specific safety items or devices (safety components).

This means that the safety of persons is achieved through machinery safety design to the relevant standard in the ANSI B11 series whilst the safety devices used in the application will need to comply with the relevant UL standards. For machines or applications where no formal regulations exist OHSA has introduced the 'General Duty Clause'.

The USA Technology Transfer Act 1996 requires federal agencies to use technical standards developed by consensus organizations where no relevant federal law exists.

So we can generally assume that ANSI standards for safety of machines that have become well established are likely to be enforceable by law through OHSA regulations.

2.8.3 Key OHSA regulations

Here are two basic OHSA regulations that are widely applicable and contain practices that are used for guidance on safety applications.

1. *CFR 1910-29-212:* Code of Federal Regulations (CFR) 29 – part 1910, Section 212. General Requirements for All Machines. Requires:

   - Appropriate safeguarding to be properly used on all machines
   - Safeguarding devices to conform to appropriate standards.

2. *CFR 1910-29-217:* Guarding of Mechanical Presses. Whilst designed for presses, this regulation is used as a reference tool for applying safety to many types of machines.

These regulations include requirements for various types of guarding systems for machines including presence sensing device initiation (PSDI).

OSHA regulations also set down the requirements for certain safety systems to be validated by the end user (employer) as well as validation by a third party who shall be an 'OSHA-recognized third-party validation organization'. Such validations are to be carried out every year and whenever a major design change is implemented on the machine. Records of such validations are to be kept for inspection by OSHA representatives. This provides the foundation for an effective policing system against gradual decline of safety standards within a factory.

There is no requirement for risk assessment for machine types as required by EU regulations but there is a separate OHSA requirement for 'Job Hazard Analysis'. This states: '*It is the duty of every employer to assess the nature of every job and ensure that workers are not exposed to unnecessary hazards*'.

2.8.4 Application standards

In addition to the standards incorporated in OHSA 29-1910, other application standards are as follows.

ANSI B11 series, specific for each type of machine application. These are considered to be the best place to start for anyone concerned with application of a machine in the USA. For example:

- B11.1   Machine tools and power presses
- B11.2   Hydraulic power presses
- B11.3   Power press brakes
- B11.10 Metal sawing machines
- B11.17 Horizontal extrusion presses
- B11.19 Safeguarding when referenced by other B11 standards
- B11.20 Manufacturing systems/cells.

The full list can be referenced via the website www.ansi.com.
For the general electrical design requirements of industrial machinery:

- National Fire Protection Association: NFPA 79.

For industrial robots the Robotic Industries Association provides:

- ANSI/RIA R15.06: Safety Requirements for Industrial Robots.

2.8.5　Design standards for safety devices

Compared with Europe there are few US design standards for safety devices. The following UL standards are available:

- UL 441　Power-operated machine control systems
- UL 508　Industrial control equipment
- UL 991　Tests for safety-related control employing solid-state devices
- UL1998　Standard for safety-related software.

2.8.6　Control reliability: foundation of US safety systems

The concept of control reliability is a common theme in the US requirements for safety controls. We shall deal with this further in Chapter 4 but at this point it is best to outline the concept because it is a basic requirement of the US safety standards. It is also very similar to the European standard for category 4 safety systems. Control reliability is defined in ANSI B11.19 as 'a method of ensuring the integrity of the performance of guards, devices or control systems'.

ANSI B11.19 – 1990 Section 5.5.1 Control Reliability
When required by the performance requirements of the safeguarding, the device or interface shall be designed, constructed and installed so that a single component within the device, interface or system shall not prevent the normal stopping action from taking place, but shall prevent a successive machine cycle. This requirement does not apply to those components whose function does not affect the safe operation of the machine tool.

The effect of the above clause is that the two main components of control reliability are fault tolerance and self-checking.

1. The only effective way to achieve fault tolerance is through redundancy in the equipment design. Through redundancy a single failure in the safety equipment cannot prevent the safe stopping of the machine.
2. Self-checking reacts to a safety system fault by preventing the guarded machine from starting until the fault has been corrected.

Note that a fault that arises during the running of the machine is not required to stop the machine, but acts to stop it being started again.

The principle of control reliability means that in the USA all machine safety functions are engineered with at least two channels of equipment in all stages of the SRECS. We shall be looking at this in Chapter 4 but the general principle can be seen in Figure 2.14.

The principles of control reliability extend into all safety applications in the USA and include the design of programmable safety systems. Effectively any device in safety system that performs a critical function must be arranged in at least a redundant pair. This is the same principle that is deployed in the European practice for categories 3 and 4 safety systems. Hence most US safety system solutions will be likely to meet EU safety requirements at the general level. Deviations will be likely in the detail levels including the electrical design practices.

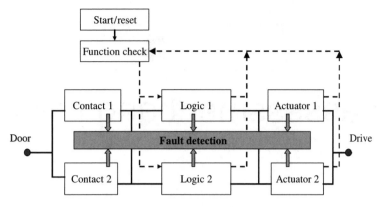

Redundancy at each stage + self-checking and start prevention

Figure 2.14
USA: control reliability example

2.9 Conclusions

- European regulations are aligned to the EU Machinery Directive for Design and Manufacture of Machines that are safe to use.
- European employers must see that their plants are safe for workers as required by the Use of Work Equipment Directives.
- EN harmonized standards support compliance with the Directives but are not integral to the regulations.
- USA suppliers and employers are required to comply with OHSA regulations and/or ANSI and other standards, which specify safety requirements.
- Safety performance standards required by both USA and EU are very similar but are not identical.
- Certification bodies can assist with compliance issues for both applications and devices.

References

OSHA 3071: Job Hazard Analysis, US Department of Labor, Occupational Safety and Health Administration.

3

Risk assessment and risk reduction

Contents summary

- Risk assessment procedure
- Typical hazards of machinery
- Systematic hazard study methods
- Risk estimation and risk ranking methods
- Risk reduction by design and by safeguarding
- Safety functions provided by control systems
- Practical exercise 2. Risk assessment example based on EN 1050.

3.1 Purpose and objectives

The purpose of this module is to develop the ability to check for hazards, conduct risk assessments and determine the risk reduction requirements for any machinery problem.

In many cases it may seem obvious that a hazard exists and that a simple solution can be put in place but we need to be aware that the severity of the consequence and the frequency of exposure will affect the extent and cost of our safety solutions.

So we must have some skills and methods available to produce consistent results for safety solutions. For this we need to:

- Know the concepts of risk, risk reduction and tolerable risk
- Have a checklist of common hazards
- Be able to select and apply some hazard study methods
- Know the basic types of risk reduction
- Know the principles laid down in EN 292 and EN 1050
- Look at a practical application.

3.2 Introduction to risk assessment

In Chapter 1 we introduced the design principle for risk reduction based on the principle of comparing an identified risk with what is known as an acceptable or tolerable risk. In this chapter we shall look at how to identify the hazards and evaluate the risks before and after adding safety measures. The result of adding protection systems such as guards and interlocks will be a reduction in risk. The overall process is summarized in Figure 3.1.

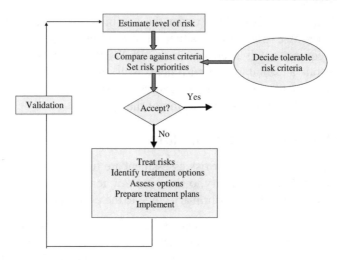

Figure 3.1
The process of risk assessment and risk reduction

Also in Chapter 1 we established the definitions or hazards and risks, so it should be clear that risk estimation requires the basic steps of:

- Identification of hazards
- Estimation of frequency of event and the consequence.

To obtain a risk estimate we need to employ a systematic method of identifying hazards and risks. This is achieved by using various methods of hazard study and hazard analysis. In particular, we need to have a comparative scale of risks either in qualitative terms such as 'high' and 'low' or in quantitative terms such as 'number of serious injuries per 1 million hours worked' or 'probability of loss of life per person'.

3.3 Procedure for risk assessment

In this section we take a look at the well-established procedures for carrying out a risk assessment on a machine or assembled group of machines. The general procedures we discuss here are based on those set down in standard EN 1050 and this in turn makes extensive cross-references to EN 292-1 and EN 292-2. This procedure forms the basis of most safety design studies that have to be carried out on machines to satisfy the requirements of the regulations. First some key points:

- Risk assessment should be based on a clear understanding of the machine limits and its functions.
- A systematic approach is essential to ensure a thorough job.
- The whole process of risk assessment must be documented for control of the work and to provide a traceable record for checking by other parties.

3.3.1 Who should do a risk assessment?

The risks from a machine are to be assessed at the design stage by the manufacturer and at the application stage by the end user.

The design team will have to carry out a risk assessment for the intended uses of the machine as part of the route to compliance with regulations. In the European case it will be done to meet the requirements of the EC Machinery Directive (see Chapter 2).

The end user is obliged under the use of work equipment regulations to carry out his own risk assessment in the situation of final use of the machine. In the UK, for example, this requirement is spelt out in the PUWER clauses where the employer is required to carry out a 'suitable and sufficient assessment' of the risks to employees. This task is naturally made easier if the designers and suppliers have already done a thorough job on the basic design and safeguarding of the machine.

3.3.2   Risk assessment procedure

The general principles for risk assessment are fully and clearly set out in the type A standard: EN 1050. Here we will try to show the key features but readers are recommended to obtain and use EN 1050 as their primary guide to the procedures.

- There are four stages to the risk assessment but these are usually followed by another stage which where risk reduction is decided and implemented. Hence we get back to the procedure described by HSE as 'five steps to risk assessment' that we saw in Chapter 1.

This procedure is illustrated in Figure 3.2 which is based on the version drawn as Figure 3.1 in EN 1050.

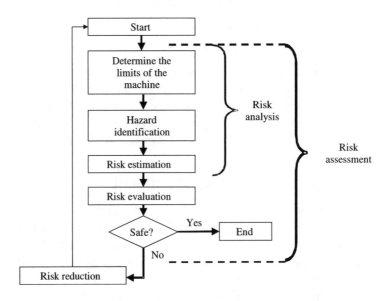

Figure 3.2
Iterative process to achieve safety (based on EN 1050)

EN 1050 provides outline guidance on the first four steps shown in Figure 3.1. The fifth step is the subject of EN 292, Part 2. The steps are explained in EN 1050 and here we provide a table to summarize the activities described in the standard:

Step	EN 1050 Reference	Activity
Define the limits of machine and its operational use	Clause 5	Take into account The phases of the machinery life Intended use and reasonably foreseeable misuse

Step	EN 1050 Reference	Activity
Obtain information for the risk assessment		Consequences of misuse Persons likely to use the machine (trained, untrained, domestic, etc.)
Identify all hazards	Clause 6	Identify all hazards, hazardous situations and hazardous events Annex A of EN 1050 lists 37 hazard types for guidance
Estimate the risks	Clause 7	Estimate the risk for each individual hazard The elements of risk are to be based on Clause 7.2. Factors affecting these elements are described in Clause 7.3. Also take into account the information for use of the machine
Evaluate the risks	Clause 8	Determine if risk reduction is required or whether safety has been achieved Decide what safety measures are needed and re-evaluate the risks after the measures have been taken into account Safety measures are achieved by: (a) Design or substitution of materials (b) Safeguarding
Modify concept or design until risks are acceptable		Note that the last step involving risk reduction is not part of EN 1050 but is the subject covered by EN 292, Part 2

Let's take closer look at some of the points raised in the table of risk assessment activities.

3.3.3 Determine the limits of the machine

The risk assessment is to be done by one or more persons. They need to have the best possible information about the machine and what is likely to be happening to it throughout its life. The scope of design activities and then the intended use including possible misuse are described in EN 292-1, Clauses 3.11 and 3.12.

These clauses remind us that the intended use must be extended to include foreseeable misuse as follows:

- Intended use of machine in compliance with the instructions
- Incorrect behavior resulting from normal carelessness
- Reflex behavior resulting from malfunctions or incidents during normal use
- Behavior resulting from taking 'the line of least resistance'
- Foreseeable behavior of non-professional users and other persons such as children.

Obviously, the proper use of the machine during its operation phase and its proper maintenance as intended by designer also play a vital role in keeping the machine safe. Thus understanding the 'intended use' and the 'limits of the machine' are equally important for the end users as well as for the designers and manufacturers. It is of course problematic to depend on users to read and follow the instruction manual!

3.3.4   Example of intended and foreseeable misuse of a machine

A very simple and everyday example of a threshing machine illustrates the above (see Figure 3.3).

A thresher being used in rural area

Figure 3.3
Threshing machine in use

The following is brief specification of the above:

> *Intended use:* The TC800 thresher-cleaner is a small, portable axial flow thresher with a cleaner. The cleaner consists of a two-directional air blast which improves the cleaning process.
>
> *Space or weight limit:* Four persons using carrying poles can carry to and from the field the machine.
>
> *Operation and maintenance (spares) requirement:* This is a simple straight-forward design, it is easy to operate and maintain (see Figure 3.4). It is a 'throw-in type' threshing combined with single oscillating screen and a winnower cleaning system.
>
> *The performance limits of the machine are:* To thresh up to 1000 kg per hour when threshing stripper-harvested paddy or up to 500 kg per hour with reaped paddy. Can thresh rice and wheat, and can shell corn.

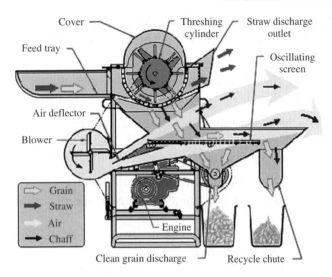

Figure 3.4
Schematic and GA diagram of the thresher-cleaner

Even in such a simple machine it can be seen if 'reaped paddy' of more than 500 kg per hour is pushed in, then it is likely to choke the thresher. This may require manual intervention of opening of cover on threshing cylinder. This will be laid the seeds for accident, which may or may not lead to injury. Logically to troubleshoot and do fault-finding for setting right the problem of 'thresher stalling', the following sequence of operation is likely to be adopted:

1. Thresher gets stalled.
2. Switch-off the engine.
3. Remove the cover of thresher.
4. Remove the over-feed.
5. Check manually for freeness of the thresher shaft.
6. Close the thresher cover.
7. Switch-on the engine.
8. Restart the threshing operation.

As 4–5 persons are generally around the machine with two persons pushing the paddy in the feed tray, anyone or two persons can intervene to implement the above actions.

The team being used to above scenario suddenly in one operation finds that even after the above the thresher is not rotating. The village mechanic is called in and he finds that the belt driving the blower has got loosened due to repeated 'overloading'. He tightens the belt and again the thresher starts working.

But this has sown the seeds of the idea in one of the bright brains for 'checking whether the thresher shaft moves before closing the thresher cover at step-6' by running the engine.

This single act could be responsible for accident and resultant injury, because at any given instance there could be a miscommunication in the team leading to accident. For instance, they start checking the belt or thresher shaft without stopping the engine or second man can start the machine when first one is still physically checking either the thresher or guard, thus causing injury to first man's hand.

The above scenario can be repeated if user puts the machine to thresh much harder stuff or uses a sub-standard belt driving the blower as replacement (due to general location of being used in a far-flung rural area).

Hence, it can be seen that even for such a simple machine how important it is for the designer and manufacturer to convey to all concerned and especially for the end users to understand:

(a) 'Intended use' – threshing of paddy only
(b) 'Limits of the machine' – only 500 kg per hour to avoid chocking.

As has been illustrated just by either not being aware of proper usage or intended misuse (for higher productivity) can give rise to unnecessary hazards and risks.

So why is it important to be aware of the limits of the machine? Because:

- It is essential that other people will know if they are working the machine outside of the range of use or scope of equipment that was the basis of the study.
- The study team must be aware of the full scope of the situations they are expected to cover.
- The records must show what scope of machine usage was considered.

3.3.5   Hazard identification

Hazards, hazardous situations and hazardous events are to be identified for all circumstances of use as defined by the previous step. This requirement in EN 1050 looks simple but it refers to the details provided in Clause 4 of EN 292-1. There we find an overview description of the hazards generated by machinery arranged into useful groups. Further details for each hazard group are defined in Annex I of the standard.

Let's take brief run through the group descriptions. What parts of a machine do you look at to find the danger areas? Typically, on a cursory glance the following areas on machinery being dangerous and strikes us that they can be a risk to anyone near the machine:

1. Parts which move or transmit power, for e.g.

   - Belts and pulleys
   - Flywheels and gear wheels
   - Shafts and spindles
   - Slides and cams
   - Chain and sprocket gear.

2. Parts that do the work, for e.g.

   - Drills and chucks
   - Tools and dies
   - Guillotine blades
   - Milling cutters
   - Circular saws.

One may question why generally only mechanical hazards pop-up in our mind? Reason is very simple, because we usually in our day-to-day life come across injuries connected with moving machinery that are derived from human contact with dangerous moving parts.

But if we give a closer look then there are a number of hazards due to other factors and these are listed hereunder. Thus hazards can be:

- Mechanical in nature
- Electrical in nature

- Thermal (generally high temperature) in nature
- Noise-related
- Radiation-related
- Material - and substance-related
- Related to bad ergonomics.

Also it could be a combination of any of the above. The following sections have more detail. Appendix A has a more exhaustive list of various hazards as listed in standards.

3.3.6   Mechanical hazards

The mechanical hazards owe their origin to human coming in direct contact with moving machinery, or a result of the impact of, entrapment by, or entanglement with plant or equipment. To understand this better we have understand movement of machinery which can be classified under four headings:

- Rotary movement
- Sliding movement
- Reciprocating movement
- A combination of the above.

Machine parts or workpieces present hazards such as:

- Dangerous shapes, bad locations, unstable masses that may topple, high-energy moving parts that may fly off.
- Energy accumulators such as springs, high-pressure liquids and gases, vacuum effects.
- High kinetic energy such as flywheels, high acceleration and velocity.
- Inadequate strength of material of construction.

Moving parts of machines while executing any of the above described motion may present a range of hazards such as:

- A straight fall into or strike or contact with moving machine part – cut and laceration injury. Machine type being – band saws, circular saws, abrasive wheels, fan blades, gear cog teeth, etc.
- Contact by being drawn or pulled into an in-running nip – crushing or severance injury from two contra-rotating parts. Machine type being – product rolls, intermeshing gear cogs, feed rollers, worm gearing, rack and pinion gearing, etc.
- Contact by being crushed – traversing movement of a machine or its part against parts of body and some fixed point such as a wall or bench or floor. Machine type – traveling tables, lathes, counter-weight blocks, gravity take ups, power presses, drop forge heads, traverse ram pushers and hydraulic doors.
- Contact with two elements of moving machinery – body part may be torn off or sheared off. Machine types – guillotines, fan blades, spoked wheels, connecting rods and linkages, oscillating tables, transfer and turnover equipment, press shears, hydraulic devices running in tight clearances, etc.
- Contact with non-cutting edges of rotating machinery – puncture or penetration of skin, friction burns and abrasions. Machine type – conveyor belts, flywheels, lips or rims of rotating drums, drive shafts and moving ropes, etc.
- Contact with punching or hole-making machinery – puncture or penetration of skin caused by flying objects. Machine type – staples, fastening pins and nails, sewing machine, drilling or punching heads, cartridge tools, etc.

- Impact to the body – Contact with robotic-type machinery while rotating, traversing, lifting or oscillating. Machine type – robotic machinery, moving tables, overhead runway equipment, hydraulic arms, etc.
- High-pressure fluid ejection – Contact with high-pressure water, steam, oil. Machine type – steam turbines, boiler feed pumps, etc.

(Ref. clause 4.2.1, 4.2.3 of EN 292-1)

To help identify some of the above hazards posed by dangerous machine parts, we may look for:

- 'Drawing in' points
- Shear points
- Impact and crushing areas
- Cutting areas
- Entanglement areas
- Stabbing points
- Abrasion areas
- Flying particles
- Any protrusions which could cause injury.

Once having identified such parts, control can be exercised.

3.3.7   Electrical hazards

(Ref. clause 4.3 of EN 292-1)

The hazards due to electrical causes are termed as electrical hazards. These hazards are the cause of danger leading to injury or sometime death. These can be classified as due to:

- Electric shock
- Electric burn
- Electric explosion or arcing
- Fire or explosion initiated by electrical energy.

Coming in contact of person with live electrical power due to the part being open (say for maintenance) or damage or deterioration of insulation, or part becoming live under fault conditions may lead to electrical shock or burn. Sometimes a small shock can also trigger off a fatal fall or a serious injury from machinery or substances being used at the time.

Electrical explosion or arcing involves considerable evolution of electrical energy, which may cause localized melting of metal, and spattered metal or radiant heat.

Fire and explosions of an electrical origin are another significant cause of electrical hazards. This could be caused due to overloading of wires, cables or electrical equipment with currents beyond their design capacity. Ignition can also be initiated by sparking, short-circuiting or arcing.

3.3.8   Thermal hazards

(Ref. clause 4.4 of EN 292-1)

These hazards are due to abnormally high or low working temperatures of surfaces exposed to human touch. Whenever the temperatures are high (say beyond 50–60 °C) any contact by touch or exposure to flame or explosions can result in burn or scalding of exposed skin. Similarly extreme cold environment may also lead to health-damaging effects.

3.3.9 Noise hazards

(Ref. clause 4.5 of EN 292-1)

The hazards associated with noise are capable of causing injury like permanent loss of hearing acuteness, tinnitus, tiredness, stress, loss of balance, loss of awareness, interference with speech communication, etc. Generally decibels of noise levels are prescribed (say 80 dB depending on national standards) and in case machines cannot be designed for it then ear-protection plugs are supplied to people associated with working of such machines.

3.3.10 Vibration hazards

(Ref. clause 4.6 of EN 292-1)

The vibrations induced either of low amplitude for a prolonged period of time or of severe in nature for short duration have potential to cause health vascular disorders like white finger, neurological, osteo-articular disorders, lumbago, sciatica, etc.

3.3.11 Radiation hazards

(Ref. clause 4.7 of EN 292-1)

These hazards are produced by non-ionizing or ionizing radiations of type:

- Low frequency
- Radio frequency and microwaves
- Infrared
- Visible light
- Ultraviolet
- X-and y-rays
- a-, - Rays; electron or ion beam
- Neutrons.

3.3.12 Hazards from materials and substances

(Ref. clause 4.8 of EN 292-1)

These types of hazards include by-products or waste materials associated with the machinery operation, or contact with materials, product or waste being ejected from machinery. These could be:

- Due to contact with or inhalation of fluids, gases, mists, fumes and dusts, having a harmful, toxic, corrosive and/or irritant effect
- Biological and microbiological hazards.

3.3.13 Hazards due to neglecting ergonomic principles

(Ref. clause 4.9 of EN 292-1)

The hazards caused due to mismatch in the machine characteristics and human abilities is likely to manifest itself in terms of:

- Human errors
- Physiological effects on account of excessive and repetitive efforts, unhealthy posture, etc.
- Psycho-physiological effects generated by mental overload or under-load, stress, etc. arising from the operation, supervision or maintenance of a machine within the limits of its intended use.

3.3.14   Hazard combinations

(Ref. clause 4.10 of EN 292-1)

Some individual hazards that seem to be minor may, when combined with each other, be equivalent to a major hazard.

The list of potential hazards can be made into a checklist that can be used by the hazard study participants. The checklist would be based on the details we have listed in Appendix A but it is important to note that the correct interpretation of the items found in the list should be made by direct reference to the relevant descriptions in EN standards listed in Annex A of EN 292-2.

3.4   Hazard study methods

The primary purpose of hazard identification is that all possible hazards are found and none are missed. This demands a structured, systematic process to identify all possible sources of harm to a person without regard being taken to the probability of harm actually occurring. This is facilitated by the use of more than one approach so that the machine and the operations associated with it are viewed from different perspectives.

There are many different adaptations of hazard study methods that have been deployed by machinery builders. It is likely that each design company will have its own preferred versions. The common theme is:

- Begin with a structuring of the machine into operations and equipment systems
- Examine each operation against checklists of possible hazards and errors
- Record all hazards and possible causes and possible consequences
- Screen out hazards and risks that are fully covered by existing safeguards and record why this has been done
- Proceed with risk estimation on those hazards that are not eliminated by the screening.

In EN 1050, Annex B, there are descriptions of several techniques for hazard analysis. The notes there make an important distinction between two basic approaches. These are called deductive and inductive. This is how the standard describes them:

In the deductive method the final event is assumed and the events that could cause this final event are then sought.

In the inductive method the failure of a component is assumed. The subsequent analysis identifies the events, which this failure could cause.

Deductive method

A good example of a deductive method is:

- Fault tree analysis or FTA. Note: IEC 61025 Fault Tree Analysis (FTA).

We have an introduction to FTA in Appendix A to this module. The technique begins with a *top event* that would normally be a hazardous event. Then all combinations of individual failures or actions that can lead to the event are mapped out in a fault tree. This provides a valuable method of showing all possibilities in one diagram and allows the probabilities of the event to be estimated. As our practical will show, this also allows us to evaluate the beneficial effects of a protection measure.

Inductive method

The so -called 'what if ' methods are inductive because the questions are formulated and answered to evaluate the effects of component failures or procedural errors in the creation of hazards at the machine. For example: 'What if the grinding wheel splits into pieces'?

This category includes:

- Failure mode and effects analysis or FMEA. Note: IEC 60812 'Analysis techniques for system reliability – Procedure for failure modes and effects analysis (FMEA)'
- Hazard and operability studies (Hazop studies). Note: IEC 61822 'Hazop studies – Application guide'
- Machinery concept hazard analysis (MHCA) described in a draft for IEC 62061.

3.4.1   Machinery concept hazard analysis (MHCA)

This procedure can generally be used as a starting point to most hazard studies. It is a process whereby a team of persons who are knowledgeable about the technical details of the machine, its modes of operation, and where and how it is intended to be used, is formed, briefed and carries out the analysis as indicated below.

Preparing for the analysis

The first stage of the process is to define the limits of the machinery as described earlier in this module.

List of hazards

The study must be provided with a prompting sheet based on the hazards list previously described. It is likely that the list will be highlighted with the particular hazards considered feasible for this machine. For example, if there are no possible sources of radiation in the machine this item could be crossed off the list at the start.

Analysis process

The functional parts of the machine should be identified, described and documented by using, for example, the categories listed below:

- Power supplies (primary and secondary)
- Controls
- Mechanisms
- Machine frame
- Others.

The most useful document here is likely to be a simple block diagram of the machine (see Figure 3.5).

The phases and modes of use of the machine on which analysis needs to be carried out should be identified and listed. These will include (see also EN 292-1, 3.11):

- Setting (or adjustment)
- Teaching/programing
- Process changeover
- Operation (this includes start-up and shutdown of the machine)

- Cleaning
- Maintenance
- Fault finding.

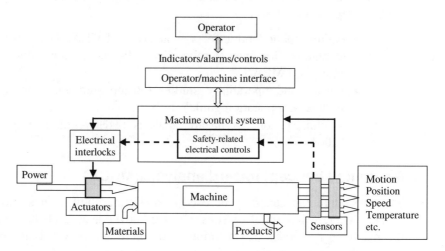

Figure 3.5
General schematic of a machine (used to identify operations and functions)

Activity analysis

Each human activity associated with each mode of use should be identified; in particular, the presence of a person(s) during each mode of use of the machine. Activities and interventions associated with non-routine modes of use (e.g. fault finding) need to be identified. It is important to determine the probable sequence(s) of activities, including misuse, that could be carried out in order to resolve an intermittent malfunction within, e.g. a sensor, control system, power control element, other operative part, on a machine.

Experience shows that hazards associated with routine operation and maintenance are more easily identified and dealt with than those associated with non-routine interventions. *There have been a significant number of accidents associated with machines, particularly automated machines, during diagnostic work or corrective actions.*

3.4.2   Results of the activity analysis

It is likely that most significant hazards will be identified as the activity analysis proceeds. Each hazardous event will be recorded in a numbered list. As a further prompt for the study team to identify hazards the following human error analysis method also described in draft IEC 62061 should be used.

Hazardous human error analysis (HHEA)   This stage of the analysis requires the team to identify the effect of human behavior on the risk arising from those activities at the machine involving persons. The tasks of the persons for all phases and modes of use of the machine should be identified and documented. The effect of human error on the occurrence of hazards and the resulting change in risk should then be estimated and the results documented. It is useful here to have a simple list of human error classifications such as those given in the following list (source: Draft IEC 62061):

Error types

- *Error of omission:*   Failure to perform an action, absence of response
- *Error of time:*   Action performed but not at or within proper time
- *Extraneous act:*   Unnecessary action not required by procedure or training
- *Transposition:*   Correct action on wrong unit, system, train or component
- *Error of selection:*   Incorrect selection control
- *Error of sequence:*   Performance of correct actions in wrong order if this is significant for success of the task
- *Miscommunication:*   Failure to communicate or receive information correctly
- *Qualitative errors:*   By excess or default (perform action incompletely).

The results for each error should have had the following queries answered and taken into consideration:

- Hazard(s) to which the operator or any bystanders would be exposed to, on account of the human error
- Range of consequences, from most usual to worst, likely to result from the hazard occurring
- External and internal factors (w.r.t. machine) that could increase or decrease the likelihood of the error occurring
- Actions/factors that could increase the risk of harm
- Actions/factors that could decrease the risk of harm, including existing safeguards which will protect against the error being made, and consequent hazard thus exposed causing harm
- Suggested safeguards required to protect against the error being made or the hazard thus exposed causing harm
- Any other actions that need to be carried out, and assigning responsibility, e.g. maintenance crew or operator or weekly inspection team, etc.

3.4.3   Hazop studies

Hazard and operability studies provide another useful approach to hazard identification and risk estimation. We have provided a brief outline of the technique in Appendix B to this module. As with the MHCA studies, Hazop studies begin with a structuring of the manufacturing process into operational stages or activities. 'What if' questions are applied in a systematic manner using standard promptings from a checklist.

The Hazop method has a good way of generating the 'what if' questions. For each operation that should be performed either by the machine (i.e. the process) or by a person operating or servicing the machine, the Hazop requires us to consider deviations from the intended operation. Here is the table of basic guidewords to be considered:

Guideword	Meaning
No or not	Complete negation of the design intent
More	Increase
Less	Decrease
As well as	Something else is done as well as the intended action. Modification
Part of	Less than a complete operation
Reverse	Opposite of the design intent
Other than	Complete substitution

Here are some basic guidewords relating to time of an action or the order in a sequence:

Guideword	Meaning
Early	Relative to the clock time
Late	Relative to the clock time
Before	Order of sequence
After	Order of sequence

The guidewords are applied to the intended operation and the study team decides if they are applicable or feasible. For example, in a machine filling beer bottles under pressure if we are looking at the system where a bottle is positioned in the filling table and a nozzle is applied to pump in the beer:

Guideword	Deviation	Cause	Consequence
More	More bottles	Sequence error Mechanical fault	Bottle jam Bottles fly out
	More beer	Feed control or sensor fault	Overfill
	More pressure	Gas supply fault	Too gassy Bottles burst
Less	Less beer	Metering error Leak or misaligned filler	Underfilled bottles Angry customers Spillage
	Less pressure	Gas supply fault Leak at filler	Flat beer
No or not	No bottle	Feeder error No bottles	Production loss

	No beer or no
	    pressure
		Blockage
		Empty tank
		Pump stopped
			Empty bottle capped
			    and packaged
			Production loss

The list of variations is rigorously tested by the study team. Hazop studies were originally developed to assist the chemical process industry to identify hazards and operational problems that will arise from deviations from the intended actions in a process plant. They have now been extended to cover a wide range of operational activities including manufacturing processes and machines. The recently issued standard IEC 61882 provides a straightforward guide to the application of hazard studies and gives several examples.

3.4.4   When to use Hazop studies

Hazop studies are good for finding weaknesses in systems (existing or proposed) where there is a movement of materials or of people or a planned sequence of activities. Hazop concentrate on deviations in the systematic operations and do not initially worry about the causes. All you have to do in the first steps is decide if the deviation is possible. So they are good for studying the performance of an assembly line or a group of co-coordinated machines in, say, a conveyor and packaging system.

Once a Hazop has identified a possible deviation, it proceeds in two directions:

1. It searches for the cause of the deviation
2. It tries to deduce the consequences of the deviation.

Hence, it quickly records the overall problem and often points to the need for a more detailed study where a problem is found to be critical.

3.4.5 Outcomes of the hazard studies

Whichever techniques are used to evaluate the specific hazards of a machine, the outcome we are expecting is a numbered list of hazards with a description of their cause and possible consequences. This provides the basis for the next stage, which is risk estimation.

We would like our documented results to look something like this:

Assembly Line No. 1: Parts assembly stage 4
Hazard identity number: AS1/4/01

Function	Robot Arm Inserts Part A into Part B to Build Assembled Part C
Problem	Sometimes assembled part C jams in the jig due to mismatching of parts A and B
Manual operation	Operator must remove jammed part C from the jig. Robot arm action must be suspended whilst operator is within range of the arm. Assembly line must be stopped but should not be powered down for this task due to complex sequencing
Hazards	Impact and crushing hazard from the robot arm Trapping of clothes and limbs due to loaded conveyor Burning due to very hot components
Consequences	Injuries range from minor to severe disabling
Causes of the event	Failure to operate 'controlled stop/hold' controls Failure to wait for robot arm to park and/or for conveyor to stop Unscheduled restart of robot or conveyor due to control faults or another operator action Failure to wear protective clothing
Action required	Risk estimate and protective measures to be provided

The hazard identification number must be designed to allow this hazard to be traceable in documentation throughout the life cycle of the plant and its machines.

3.5 Risk estimation

Now that we have a stock of methods for identifying the hazards we are ready to rate the risks they present to humans and to the machinery or the business that uses them. The basics of risk estimation are spelt out in EN 1050 in paragraphs 7.1 to 7.3.8. The principle of risk estimation is to combine the elements of consequence of the accident with the likelihood of it happening. The likelihood is based on the elements of frequency and duration of exposure combined with the probability of the accident occurring whilst the person is exposed to the hazard and the possibilities of avoiding or limiting the harm (see Figure 3.6).

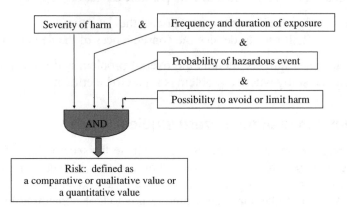

Figure 3.6
Risk estimation principles and elements

3.5.1   Elements of risk

It's important to understand the factors to be considered for each element of risk. These are described very well in EN 1050 and our notes here are based on the standard.

- *Severity (degree of possible harm):*   The severity of the harm is a function of the objective of protection (i.e. human, property or environment) in relation with the severity, i.e., slight, serious or fatal and it does affect one person or several. Harm may also occur to property or the environment.
- *Probability of occurrence of harm:* The probability or likelihood of occurrence of harm can be estimated by taking into account the frequency and duration of exposure, probability of occurrence of a hazardous event and the possibility of avoiding or limiting harm.
- *The frequency and duration of exposure:*   This is influenced by the need for access to the danger area by persons, the duration, number of persons and frequency of access into the hazardous area of machine for maintenance or during normal operation.
- *The probability of occurrence of a hazardous event:*   This has to be estimated by considering historical data on the history of similar accidents and/or calculating the rate of failures of equipment based on reliability analysis or failure rate data. Where the primary cause is human error there are tables of human error rates that can be used to assist with estimates.
- *The possibility of avoiding or limiting harm:*   This is influenced by a number of factors including:

    - The type of persons involved and their practical experience in operating the machinery
    - The speed at which the hazard occurs and the extent to which there can be any early warnings
    - A person's awareness of the hazard through its nature or through information or warning signs
    - The person's ability to avoid the hazard by escape or reflexes.

3.5.2   Aspects to be considered when establishing elements of risk

The following need to be given due credence while estimating the elements of risk:

- *Persons exposed:* This shall include not only operators but also any other person likely to be present in the vicinity or in harms way of machine when the hazard is likely to occur.
- *Type, duration and frequency of exposure:* This shall include the influence of necessity of exposure of persons to hazards for all modes of working of machinery like setting, testing, training, process changeover, correction, cleaning, troubleshooting and maintenance. It may also include the risk estimation for situations when safety may be defeated for maintenance purposes.
- *Relationship between exposure and effects:* The cumulative effect of exposure to the risks shall be done objectively based on the published data available from various sources.
- *Human factors:* The human factors play a vital role in controlling the effect and extent of the harm. Hence, various factors influencing the human behavior should be taken into account. Some of these are:
  - Interaction of person with the machinery
  - Interaction between persons
  - Psychological aspects
  - Ergonomic effects and its influence on design of the machinery
  - Ability of persons to perceive the risks and his awareness toward the same
  - Natural or developed ability of persons to execute the tasks in relation to the machine under review
  - Person's psychology to temptations for shortcuts and executing the tasks without deviations either intentional or un-intentional.

- *Reliability of safety function:* EN 1050 calls for the reliability of safety functions to be considered. As we shall see in Chapter 4, the type B standards for safety system design and the design of safety components provide more specific details that assist in this regard.
- *Possibility to defeat or circumvent safety measure:* The possibility of defeating a safety measure by tampering or taking shortcuts is highlighted here by EN 1050. It makes some very notable points here that there may be incentives to defeat the safety measure if it:
  - Slows down production or interferes with other preferences of the operator
  - Is difficult to use
  - Persons other than the operator are involved
  - Is not recognized by the user or is not accepted as suitable for the job.

- *Ability to maintain safety measures:* Risk estimation shall consider whether the safety measures can be maintained in the condition necessary to provide the required level of protection. This places a responsibility on the safety system designer to provide good practical and maintainable solutions.
- *Information for use:* Risk estimation shall take into account the quality of the information for use passed to the persons using the machinery. This covers warning signs, proper instructions for operation and maintenance, and the question of how these instructions reach the operatives. EN 1050 references the details on this subject contained in EN 292-2, paragraph 5.

The above range of factors provides us with a valuable checklist of points to consider when we are faced with delivering a good estimate of the risk presented by a machinery hazard.

3.5.3   Practical approaches to risk measurement

There is generally a problem with risk estimation when it comes to practical application. As we have already seen in the discussion on qualitative vs quantitative risk measurement, it is often very difficult to define a risk in simple numerical terms. But it is equally difficult to have everybody using different descriptions of risk, say 'high' or 'very high' or 'moderate'.

Risk is something we can measure approximately by creating a scale based on the product of frequency and consequence. For example, we can measure consequences in terms of injury to persons. Here is a quantitative scale:

- *Minor:*   injury to one person involving less than 3 day's absence from work
- *Major:*   injury to one person involving more than 3 day's absence from work
- Fatal consequences for one person
- *Catastrophic:*   multiple fatalities and injuries.

Likewise, the frequency or likelihood of an event causing injury can also be placed on a scale. For example, here is a qualitative scale (descriptive but does not define numbers):

- Almost impossible
- Unlikely
- Possible
- Occasionally
- Frequently
- Regularly.

Alternatively, frequency can be placed on a quantitative scale. This would simply be the event frequency in events per year. For example:

- One hazardous event occurring on the average once every 10 years will have an event frequency of 0.1 per year.
- A rate of 10^{-4} events per year means that an average interval of 10 000 years can be expected between events.

Another alternative is to use a semi-quantitative scale or band of frequencies to match up words to frequencies. For example:

- Possible = less than once in 30 years
- Occasionally = more than once in 30 years but less than once in 3 years
- Frequently = more than once in 3 years
- Regularly = several times per year.

Once we have these types of scales agreed the assessment of risk requires that for each hazard we are able to estimate both the likelihood and the consequence. For example:

- *Risk item No. 1:*   'major' injury likely to occur 'occasionally'
- *Risk item No. 2:*   'minor' injury likely to occur 'frequently'.

Whilst both the above items are undesirable we cannot yet tell which of them is the most important problem and in need of risk reduction, or even if they need any reduction at all. *What we need is a system of comparative values for risk.*

3.5.4 Introducing the risk matrix

From the above it is clear that a scale of risk can be created from the resulting products of frequency and consequence. One popular way to represent this scale is by means of a simple chart that is widely known as a risk matrix. Here are some examples.

Figure 3.7 is a simple risk matrix where frequency of the possible event is ascending on the *Y*-axis and the consequence categories are ascending on the *X*-axis.

When the product of frequency and consequence is high, the risk is obviously very high and is unacceptable. The unacceptable region extends downwards toward the acceptable region of risk as frequencies and/or consequences are reduced. The transitional region as shown in the figure is where difficult decisions have to be made between further reduction of risk and the expenditure or complexity needed to achieve it. Our figure shows some attempt at quantifying the frequency scale by showing a range of frequencies per year for each descriptive term. This is usually necessary to ensure some consistency in the understanding of terms used by the hazard analysts.

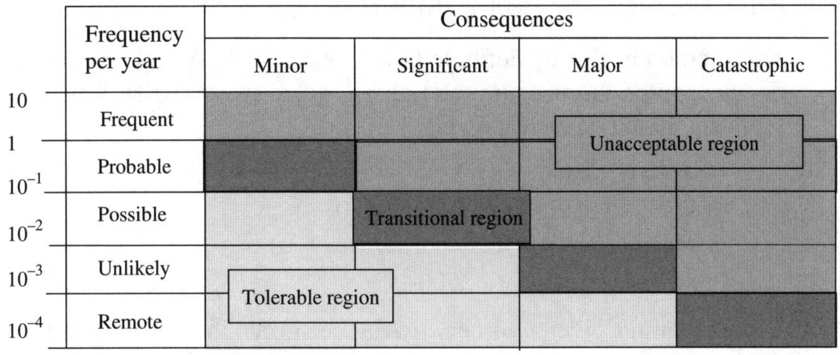

Figure 3.7
Risk matrix with tolerability bands

3.5.5 Risk ranking scales

Some companies go a step further and assign scores or values to the descriptions of frequency and consequence. This has the advantage of delivering risk ranking on a numbered scale, allowing some degree of comparison between risk options in a design. Figure 3.8 shows a possible score and ranking values on the same risk matrix as above.

The scoring system adopted in the above figure is an arbitrary scheme devised to suit the tolerability bands as best as possible. Each company and each industry sector may have its own scoring system that has been developed by experience to provide the best possible guidelines for the hazard study teams working in their industry. There does not appear to be any consensus on a universally applicable scoring system but the ground rules are clear. The scales must be proportioned to yield consistently acceptable results for a number of typical cases. Once the calibration of a given system is accepted, it will serve for the remainder of a project.

Frequency per year	Consequences			
	Minor: 1	Significant: 3	Major: 6	Catastrophic: 10
Frequent: 10	6	10	60	100
Probable: 8	8	24	48	80
Possible: 4	4	12	24	40
Unlikely: 2	2	9	12	20
Remote: 1	1	3	6	10

(Left axis scale: 10, 1, 10^{-1}, 10^{-2}, 10^{-3}, 10^{-4})

Figure 3.8
Risk matrix showing ranking and tolerability bands

3.5.6 Published risk scales for machinery practice

Some of the experienced practitioners such as the leading safety control and sensor suppliers advise what amounts to semi-qualitative scales for risk estimating. Usually these methods are specifically intended to help with the selection of protective systems. For example, we show here the PILZ system suggested in their safety manual.

3.5.7 PILZ method for risk estimation

This method begins by defining four levels of risk. A numerical weighting is given to each risk descriptor and the sum total of risk value is arrived at. This value is indicative of risk estimated based on the following:

- Negligible   0–5
- Low but significant   5–50
- High   50–500
- Unacceptable   500+

The weightings given to other criteria are shown in the following table:

Likelihood of Occurrence (LO)/Contact with Hazard	
Almost impossible – possible only under extreme circumstances	0.033
Highly unlikely – though conceivable	1.0
Unlikely – but could occur	1.5
Possible – but unusual	2.0
Even chance – could happen	5.0
Probable – not surprising	8.0
Likely – only to be expected	10.0
Certain – no doubt	15.0
Frequency of Exposure to Hazard (FE)	
Annually	0.5
Monthly	1.0
Weekly	1.5
Daily	2.5
Hourly	4.0
Constantly	5.0

Degree of Possible Harm (DPH), Taking into Account the Worst Possible Case	
Scratch/bruise	0.1
Laceration/mild ill-effect	0.5
Break minor bone or minor illness (temporary)	2.0
Break major bone or major illness (temporary)	4.0
Loss of one limb, eye, hearing loss (permanent)	6.0
Loss of two limbs, eyes (permanent)	10.0
Fatality	15.0
Number of Persons Exposed to the Hazard (NP)	
1–2 persons	1
3–7 persons	2
8–15 persons	4
16–50 persons	8
50+ persons	12

The calculation is as follows:

$$LO \times FE \times DPH \times NP = \text{Risk Level}$$

The risk assessment exercise should involve the operating level people (both workers and supervisors), should take into account the working environment, skills and other productivity-related factors into account. All hazards as given in EN 1050 and other relevant standards should be checked for and assessed.

Most importantly these need to be documented and corrective measures to control the risk initiated.

3.5.8 Guardmaster method for risk estimation

Here is the method suggested by Guardmaster, a UK-based supplier of protective devices such as machinery guards and light curtains. As with the PILZ method the supplier stresses that this is 'intended to encourage a methodical and documented structure'.

The scale of severity of an accident is suggested as follows:

- Fatal = 10
- Major involving permanent disability = 6
- Serious, involving breakages or burns or loss of consciousness but ultimately recoverable = 3
- Minor = 1.

Frequency of exposure to the hazard:

- *Frequent:*   Being several times per day = 4
- *Occasional:*   Being daily or less than daily = 2
- *Seldom:*   Being weekly or less = 1.

The probability that an injury will occur:

- Certain = 6
- Probable = 4
- Possible = 2
- Unlikely = 1.

Additional factors   Guardmaster suggests that the total of the above scores be adjusted by adding the following allowances for factors which may only be apparent when the machine is installed in its operation location.

- More than one person exposed to the hazard at the same time: multiply the severity factor by the number of people.
- If time in the danger zone exceeds 15 min, add 1 point to the frequency factor
- If operator is unskilled or untrained, add 2 points to the total.
- For very long intervals between maintenance access, double the selected frequency factor.

Add the above factors to the previous total to get an overall score.

Risk bands

- High risk is indicated if the score exceeds 15.
- Medium risk lies in the band 6–15.
- Low risk is indicated if the score falls below 6 and ideally we want to see a score around 1 or 2.

This approximate scale of risks shows the factors that contribute to high score and it allows us to quickly see if a risk is unacceptably high. It also allows us to compare one particular risk with another within the same set of value judgments used by a particular study team. It is not of course a universal scale of risk and must be treated with caution.

3.5.9   Using the scales for risk reduction

On the basis of the typical scoreboard systems shown above, it is easy to see that if risk reduction measures are proposed the score sheet can be re-evaluated for the level of risk remaining after the measures have been assumed to be in place. Since the measurements are comparative on the same scale it is not too sensitive to the absolute scale of measurement employed.

The problem of determining the risk scale and how risk reduction relates to the reliability (or more correctly the integrity) of the safety measures is a persistent one that has only recently been tackled with some success in the process industries. We shall see something of this in later modules. At present the new standard being drafted for machinery safety using programmable electronic systems incorporates a points system similar to the ones we have seen above but it is still in development. At the moment the only standard method relating to risk reduction is the simple decision chart for safety categories given in EN 954-1, which we shall be looking at in the next module.

3.5.10   Tolerable risk concepts: how to know what level of risk is acceptable?

The question of what level of risk is to be accepted is a complex one and we have added some notes on this subject in Appendix C to this manual. The standards generally call for comparisons to be made with existing similar machines in apparently safe use. In Appendix C, we list quantitative data from various sources and explain the concept of probability of harm as a statistic. This can be related to the Alarp principle we introduced in Chapter 1 but it is often difficult to use quantitative date in practice.

3.6 Risk reduction principles

Once an unacceptable level of risk has been identified the methods to be used for risk reduction will follow a well-defined path based on the principles laid down in EN 292-1. These principles are easy to see from Figure 3.9, which is based on Table 2 of EN 292-1.

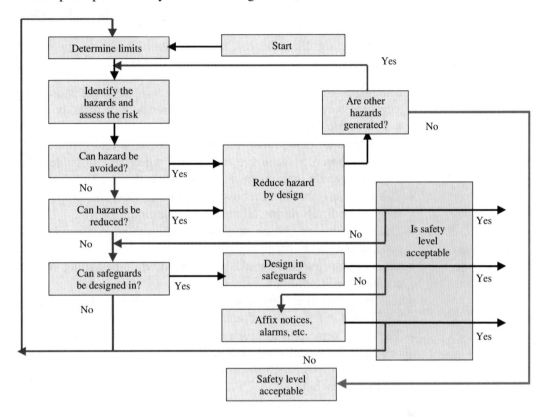

Figure 3.9
Strategy for selecting risk reduction measures (based on EN 292-1, Table 2)

The order of activities for selecting safety measures is:

- Remove the hazards or limit the risks as much as possible by design (methods defined by Clause 3 of EN 292-2).
- Provide safeguarding to protect persons from risks that cannot reasonably be eliminated or avoided by design (methods defined by Clause 4 of EN 292-2).
- Inform and warn the user against any residual risks (methods defined by Clause 5 of EN 292-2).
- Consider additional precautions (methods defined by Clause 6 of EN 292-2).

Note that the residual risk level must be checked at each stage of the process through the risk reduction exercise. If the risk is negligible after the design work has implemented its safety features, there may be no need for the next stages of safeguards and warning notices.

3.6.1 Risk reduction by design

Clause 3 of EN 292-2 has an extensive range of important design features in machines that will contribute to reducing risk. We must not risk overlooking any of the safe design principles in our rush to add on the safeguarding devices. Most importantly for us this clause includes the outlines of safe design practices for the safety-related parts of electrical control systems.

Here is a brief review of the main design features described in EN 292-2 that should be checked off before we proceed with a safeguarding plan. For a full description readers should refer directly to EN 292-2 paragraphs 3.1–3.12 where the principles of risk reduction by design are presented very effectively.

The first way of controlling the risk is intervention at the design stage by choosing appropriate design features so as to reduce as many of hazards as possible. These include:

- Minimizing the usage of sharp edges, corners, protruding parts and incorporating safe gaps between two moving parts (either no part of body is able to enter or there is sufficient gap for the part to enter).
- Using the minimum kinetic energy, limiting noise and vibration, etc.
- Using inherently safe technologies wherever possible. For example, intrinsically safe instruments in an explosive atmosphere.
- The materials should be chosen keeping in view the properties and stresses, etc., in safe limits during all modes of operation.
- The technologies chosen should be safe and in conformance with relevant codes and industry practices.
- Ergonomic principles should be observed while designing such that operators are subjected to minimum stress.
- It is very important that manual controls, displays, etc., including E-stops are engineered in a way such that high visibility is imparted to them.
- The safety principles as given in codes for designing of machine controls shall be scrupulously followed, such that deterioration in control components, power supplies, abrupt change in operating mode, etc., is not encountered during the life of the machine.
- By limiting the operator intervention in hazardous zones through selection of appropriate operating mode and control regime. Ensure safe modes of machine operation while setting, testing, tuning, training, process changeover, troubleshooting and day-to-day maintenance.
- The designer should also take appropriate steps to have full compatibility with EMC guidelines.

Key points for the workshop compliance with this part of the standard requires:

- Good mechanical design
- Use of ergonomic principles
- Good design practices for SRECS
- Safe practices for hydraulics and pneumatics
- Prevention of electrical hazards through electrical design in accordance with EN 60204 or similar codes.

It is important to be aware that the safety-related controls built into the design would include regular features such as controlled stops and interlocks. These will most probably be integrated into the basic control scheme but at this point they will be identified as safety-related.

- SRECS functions in the basic control scheme must be identified.
- Safety category of each of the SRECS will have to be decided.

- SRECS may have to be fully separated from the PLC system that operates the basic control of the plant.
- SRECS may merit the use of a separate safety PLC to be qualified under the new IEC 61508 standard (see also Chapter 9).

3.6.2 Risk reduction by safeguarding

In many cases of safety studies for a machine many of the above features will have already been built into the design and we are left with contemplating what safeguarding devices we can use to further reduce the risk. At first glance it seems that there is a bewildering array of safeguarding devices on the market but the EN 292 standard has succeeded in simplifying the choice into a small number of types. We are going to look at the characteristics machinery-protection devices in the next two modules of the workshop. For the moment we just need to see what influences the choices.

Figure 3.10 is based on Table 2 of EN 292-2 and helps us to make the choice of safeguarding method according to the hazards generated by moving parts.

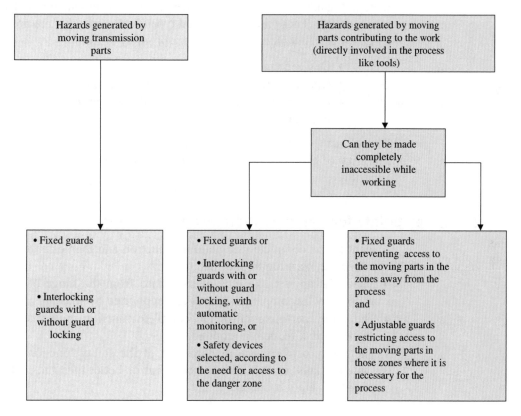

Figure 3.10
Choice of safeguarding methods

Here is a summary of the selection guidelines in Clause 4.

Clause 4.1.1 General
The guidelines in the standard are advisory only:

- Choice of guard type to be made on the basis of risk assessment
- Fixed guarding is simple but inconvenience increases as the need for access increases.

Clause 4.1.2 Access not required during normal operation
Fixed, interlocking or self-closing guards are suitable.

Clause 4.1.3 Access required during normal operation
More convenient types such as interlocking guards or even two-hand control devices are needed to improve productivity. The most effective and convenient types for rapid but safe access are usually light barrier types. The standard groups these as 'trip devices'.

Clause 4.1.4 Access required for machine setting, teaching, cleaning, etc.
Use control measures such as 'manual control mode'.

Clause 4.2 Requirements for the design and construction of guards and safety devices
This clause describes the basic requirements for design of guards. We shall look at these in the next few modules of the workshop. At this point we should note that EN 292-2 as a type A standard has set down broad guidelines that have been used by standards writers to produce specific design standards for each type of safety device. These are the so-called type C standards that are now in service.
Clause 4.2.3 Technical characteristics of safety devices

- Designed to safety principles
- Operated and connected so that they cannot be easily defeated
- Performance to be consistent with the control system to which they are integrated.

These last points are significant because they remind us that the safety integrity of the device must match the safety integrity of the complete safety function.

3.6.3   Key points for risk reduction by safeguarding

- Identify the complete safeguarding function and then establish what degree of risk reduction it has to provide.
- Select the appropriate type of safeguard from the range of devices available assisted by the simple guidelines and supported by standards for each type.
- The safety performance of the complete function must match or exceed the degree of risk reduction required.
- The safety performance of each part of the safety function (e.g. guard, limit switch, relays, contactor) must be equal or better than the performance of the complete safety function.

3.6.4   Risk reduction by information for use

Information for use is an integral part of the supply of a machine. The information will cover the purpose and range of use of the machine and where needed it must inform and warn the user about residual risks. Hence devices such as audio and visual alarms and warning signs provide a legitimate contribution to risk reduction.

However we are warned against using such measures as a cheap way out! It is not satisfactory to leave out safety guards and just put up a notice saying, 'This machine is dangerous'. It says in Clause 5.1.2 'Information for use shall not compensate for design deficiencies'.

3.6.5 Key points concerning risk reduction by information

Warnings on the machine or in the handbook: choices depend on the risk and time when the information is needed. Standardized phrases should be used in all warning statements. Signals and warning devices should be:

- Emitted before the occurrence of hazardous event
- Unambiguous
- Clearly perceived and differentiated from all other signals (see codes of practice)
- Can be clearly recognized by the users
- Easy to check and instructed for regular checking.

Designers are notified of the risk of 'sensor saturation'. We might see this as 'alarm overload' or the problem of too frequent operation of an alarm. These factors place limitations on the true value of an alarm in the risk reduction role.

A good example of this is the beeping noise made by a motorized trolley tug at the big airports such as London Heathrow. The sound is almost always present and very quickly looses its value in a busy arrivals hall.

3.6.6 Risk reduction by additional precautions

These additional precautions can be divided broadly as – first those deployed for emergency scenarios and secondly usage of appropriate equipment, systems and arrangements contributing to safety.

The first category include deployment of at least one number 'E-stopping device' with each machine, which shall stop the operation of machine from any mode. The release of this stop after being activated should be reset by appropriate operation and such release shall not restart the machine but only enable start permissive. This provision will be deviated in special cases like hand-held machines where stoppage does not reduce the risks, etc. Enough precautions should be taken so that trapped persons are rescued where trapping of operator is envisaged.

In the second category are included provisions for:

- Maintainability of machine
- Isolation and energy dissipation
- Easy and safe handling of machines and their heavy component parts
- Safe access to machinery
- Stability of machines and their elements
- Diagnostic systems to aid fault-finding and rectification.

It is important to note that isolation of machines from all power supplies shall be either visible or ensured by locking of a visible switch in safe position. The areas of machine so isolated should be clearly demarcated. The arrangements shall be made to dissipate the stored energy safely and all this to be ensured through a properly established 'work-permit' system.

Suitable diagnostic systems shall be engineered to enable quick troubleshooting so that overall availability and maintainability improves.

3.6.7 Conclusions on risk reduction

What we have seen in this review of the risk reduction principles is an example of the well-proven method of 'belt and braces'. The process industries call this the application of layers of protection. Figure 3.11 shows how each risk reduction measure provides a safety layer that contributes to overall safety.

Risk reduction measures can be seen as layers of protection

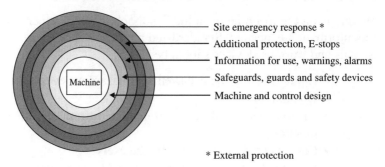

Site emergency response *
Additional protection, E-stops
Information for use, warnings, alarms
Safeguards, guards and safety devices
Machine and control design

* External protection

Figure 3.11
Layers of protection reduce the risk presented by a machinery hazard

Each of the risk reduction measures will make a contribution in proportion to its SIL. Figure 3.12 indicates three protection layers with risk being reduced in steps. At the end of the chain of measures there will be a small but acceptable residual risk.

Risk reduction layers

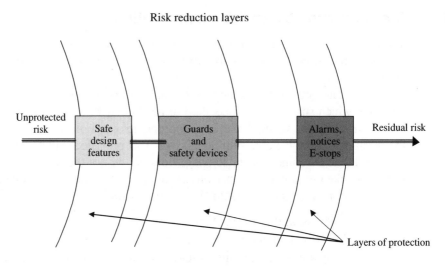

Figure 3.12
Examples of three independent protection layers reducing risk

Once the design team has proposed suitable risk reduction measures, it is required to verify that the measures are practicable and affordable. It is important at this stage to take into account factors that can affect the success of the proposed safety measures. Here are some points to consider.

- Risk assessment relies on the judgment of the assessor and hence the assessor should record the reasoning behind each decision.
- Involve operators and supervisors in the assessment. Their first-hand experience will be of great value.
- Take into account the skills and awareness of the exposed persons. The dangers of boredom and familiarity with the equipment must be considered.
- The full hazards list should be checked to cover possibilities of all types including high-temperature surfaces, fires, chemical and oil releases, explosions, electric shocks, etc.

3.7   Outcomes of the risk assessment

3.7.1   Practical example

Let us see how all of this works out for the example of the robot arm we saw earlier.
Assembly line No. 1: parts assembly stage 4

Function	Robot arm takes part A from an oven and inserts into part B on moving conveyor to build assembled part C. Conveyor transfers assembled parts into cooling tunnel
Problem	Sometimes assembled part C jams in the jig due to mismatching of parts A and B
Manual operation	Operator must remove jammed part C from the jig. Robot arm action must be suspended whilst operator is within range of the arm. Assembly line must be stopped (hold condition) but should not be powered down for this task due to complex sequencing

Hazard Item No.	Hazard	Caused by	Consequence

AS-1/4/01	Robot arm collides with person
	Impact and crushing hazard
		(a) Person too close to moving robot arm during normal production
		(b) Robot arm movement during operator action to clear jammed parts due to:
		   Failure to operate 'Controlled stop/hold' controls
		   Failure to wait for robot arm to park
		   Unscheduled re-start of robot
			Major injury to workers
			Major injury to operator

AS-1/4/02	Moving conveyor can trap clothes and limbs	Conveyor movement whilst operator is removing faulty parts due to: Failure to operate 'Controlled stop/hold' controls Failure to wait for conveyor to stop Unscheduled re-start of conveyor	Major injury to operator
AS-1/4/03	Burns hazards from heated parts on the conveyor	Contact with persons during normal production Operator attempts to handle jammed parts without gloves	Minor injury to operator

Initial risk estimate

The first step is to estimate the risk without any control measures in place. This will dimension the risk reduction facilities we have planned or will need. The estimate here is tabulated for the various factors to be considered. For comparison there are point-scoring systems for both PILZ and Guardmaster.

Risk estimate for robot arm collision with persons:

(a) Persons in the area inc. cleaners

Risk Parameter	Rating Using EN 1050	PILZ Rating	Guardmaster Rating
Degree of harm	Slight to serious	DPH = 4	Reversible = 3
Probability of harm Frequency and duration of exposure Likelihood Avoidance possibilities	5 min per hour for one person Probable Good	FE = 4 LO = 5	>1/day = 4 Probable = 4 Unskilled = 2
No. of persons	One	NP = 1	One
Risk summary	High risk of injury to one person. Unacceptable	80 High	13 Moderate

(b) Robot arm movement during operator action to clear jammed parts

Risk Parameter	Rating Using EN 1050	PILZ Rating	Guardmaster Rating
Degree of harm	Serious	DPH = 6	Permanent = 6
Probability of harm Frequency and duration of exposure Likelihood Avoidance possibilities	5 min per hour for one person Possible Poor	FE = 4 LO = 4	>1/day = 4 Possible = 2 Skilled = 0
No. of persons	One	NP = 1	One
Risk summary	High risk of injury to one person. Unacceptable	96 High	12 Medium to high

For the above two cases that are closely related the risk is unacceptable and the next step will be to decide on how the risks can be eliminated or reduced to an acceptable level. For this we have to:

- Define the 'safety requirements' as an overall objective.
- Determine a 'safety function' that satisfies both the safety requirements and still meets the overall objectives of the machine ..., i.e. still allows the machine to work properly and deliver its product at the best possible speed and efficiency.
- Define the 'functional requirements' of each part of the safety function.

The overall safety requirement can be defined as 'ensure that no persons are exposed to the moving parts of the robot arm and the conveyor under normal production conditions including the task of removing jammed parts'. To meet this 'safety requirement' we have to determine the possible solutions and define the 'functional requirements' of the safety measures.

We first have to search for inherently safe design for the assembly line but in this case we are going to assume we are stuck with having to use a moving arm robot type of device to carry out the assembly operation. We can however specify controls to assist safety.

The assembly sequence control shall have a "sequence stop" mode that parks the robot arm in safe position and stops the conveyor. The sequence cannot be restarted until the operator has reset the stop condition and then pressed the start button.

This arrangement ensures that robot parks in a safe position and that no damage is done by a sudden crash stop.

Now we consider protection measures in concept. Obviously there is scope for a fixed guard around the robot area that will prevent any person from accidentally walking into the danger area. But because the machine operator requires access from time to time to clear away the jammed parts there has to be provision for entry to the danger area under the right conditions. There is no requirement for anyone to be near the robot arm when it is moving, so we can specify the functional requirements as follows:

Part 1: A fixed guard will enclose the zone of movement of the machine to prevent a person or any part of a person from being closer than 0.5 m from the moving parts of the robot at its fullest extent. The same shall apply to the conveyor.

Part 2: Access to areas within the zone of movement shall only be possible via a door. The door cannot be opened when the contactors supplying power to the robot actuator and the conveyor drive are closed. The same contactors shall be prevented from closing unless the guard door is closed.

In this example it is clear that our concept is an electrically interlocked guard door with a safety control interlock on the door position. This will prevent access to the danger area until all movement has ceased and will not permit operations unless the safety door is shut. Note it does not safeguard the operator if someone else closes the door whilst he is still in the danger area, but since there is always line of sight visibility we may be able to accept that chance.

3.7.2 Re-assess the safety of the machine

With the first of our safety functions defined we should now revisit the risk assessment for this particular hazard and decide if the risk is now acceptable.

Risk Estimate for Robot Arm collision with persons:

(a) Persons in the area inc. cleaners

Risk Parameter	Rating Using EN 1050	PILZ Rating	Guardmaster Rating
Degree of harm	Slight to serious	DPH = 4	Reversible = 3
Probability of harm Frequency and duration of exposure Likelihood Avoidance possibilities	No access possible due to guards. Cleaning access required once per week. Power locked off by open guard door. Possible only through misuse Good	FE = 1.5 LO = 1	1/week = 2 Unlikely = 1 Unskilled = 2
No. of persons	One	NP = 1	One
Risk summary	Very low risk of injury to one person. Acceptable	6 Low	7 Low

(b) Robot arm movement during operator action to clear jammed parts

Risk Parameter	Rating Using EN 1050	PILZ Rating	Guardmaster Rating
Degree of harm	Serious	DPH = 6	Permanent = 6
Probability of harm Frequency and duration of exposure Likelihood Avoidance possibilities	5 min per hour for one person Almost impossible Poor	FE = 4 LO = 0.033	>1/day = 4 Unlikely = 1 Skilled = 0
No. of persons	One	NP = 1	One
Risk summary	Very low risk of injury to one person. Acceptable	0.8 Negligible	11 Low to medium

The results shown above show some discrepancies between the general risk assessment and the semi-qualitative methods suggested by the two vendors. The problem here is that the frequency of exposure to risk does not change for the operator whilst the likelihood of the accident has been reduced substantially. The Guardmaster scoring system does not give much credit for reducing likelihood and it does not use the principle of multiplying consequence by frequency to establish risk.

In the risk models used by IEC 61508 we would see proportional credit for reducing likelihood. It is clear that there is much room for interpretation here. This is one of the reasons why it is essential to record the reasoning behind the risk estimate at every step.

We can leave the second safety function out as an example as it will follow the same path as the first one. Lets briefly test the third safety function, this being the need to avoid burns from handling the heated components.

Risk estimate for accidental burn through handling components:

(a) Persons in the area inc. cleaners

Risk Parameter	Rating Using EN 1050	PILZ Rating	Guardmaster Rating
Degree of harm	Slight to serious burn	DPH = 2	Reversible = 3
Probability of harm Frequency and duration of exposure Likelihood Avoidance possibilities	Frequent Even chance if op forgets to wear gloves High	FE = 2.5 LO = 5	Frequent = 4 Likely = 4 Skilled = 0
No. of persons	One	NP = 1	One
Risk Summary	Moderate risk of injury to one person. Not acceptable	25 Low but significant	11 Medium

In this application the need to handle very hot components presents a moderate risk. The overall safety requirement is to prevent burns injuries to the operator. The risk reduction sequence says:

- Can we change the design to avoid this operation?
- Can we provide safeguards?
- Can we provide information for use?

If the first option is not possible we could consider a special tool, such as tongs, for handling the hot components. This may not be practicable. We could delay the release of the access door for 5 min to allow time for parts to cool, but this would impact production. This would have to be considered.

Lastly we could place warning notices at the access door and at the conveyor side. The notices will state that there is a risk of very hot surfaces and requires that any person entering the protected area must always wear heat-protective gloves and overalls.

If we assume that the warning notices will be sufficient, the record will show the reasons for this decision and the re-assessment record will be as follows:

Revised risk estimate for accidental burn through handling components

Risk Parameter	Rating Using EN 1050	PILZ Rating	Guardmaster Rating
Degree of harm	Slight to serious burn	DPH = 2	Reversible = 3
Probability of harm Frequency and duration of exposure	Frequent	FE = 2.5	Frequent = 4
Likelihood	Possible but unusual. Only if operator ignores warning	LO = 2	Possible = 2 Skilled = 0
Avoidance possibilities	High		
No. of persons	One	NP = 1	One
Risk summary	Slight risk of injury to one person. Acceptable	10 Low but significant	9 Low to moderate

In conclusion, this example shows that whilst formulas can help there are many aspects to risk assessment that depend on the experience and judgment of the persons involved. The assessment by one group of persons must be documented carefully so that anyone else returning to this record can follow the reasoning used.

3.7.3 Final evaluation

Having completed the risk assessment and specified the safety functions, the design team will have to decide if the overall results are now likely to deliver a safe machine EN 1050 provides us with Clauses 8.2 and 8.3 to assist with deciding if the risk reduction process can be concluded.

3.7.4 Achieved objectives

Summary of risk reduction objectives, which will indicate that process of risk reduction, can be concluded:

- Risk is eliminated by substitution of less hazardous materials and substances and by suitable safeguarding.
- Choosing safeguarding such that safe situations are achieved during intended use, like reducing severity of harm, probability of defeat or circumvention of danger, etc.
- Providing sufficiently clear information for intended use is furnished.

- Providing fairly clear operating and maintenance instructions appropriate for those who are going to use it and those who are likely to present in the vicinity of machine during usage or shutdown.
- Describing relevant safe working practices along with the training requirement.
- Appropriately informing the user about 'residual risks' in different phases of the life of the machinery.
- Adequately describing recommended personal protective equipment need to be provided along with training requirement if any.

3.7.5 Comparison of risk

One method of demonstrating that risk has been reduced to an acceptable level is to compare the machine in question to similar machines that are in service and are considered to be safe. Clause 8.3 of EN 1050 gives us some guidelines on this as follows:

- The similar machine is safe
- The intended use and technologies employed of both machines are comparable
- The hazards and the elements of risks are comparable
- The technical specifications are comparable
- The conditions for use are comparable.

3.8 Documentation methods for the risk assessment

We have seen in the above example that a tabulated set of records can be used for recording the risk assessment. The contents of the records should include the following points. These are set out in checklist form. The list could also be used to serve as an index to the relevant documents.

3.8.1 Checklist for risk assessment documentation

Subject	Included or N/A	Description or Document Ref. No.	Date Completed
Scope of machine assessed Description, limits			
Intended use			
Any relevant assumptions (such as loads, strengths, safety factors)			
Hazards identified			
Hazard study references			
Information basis: Data and sources Previous experiences			
Objectives of safety measures defined and recorded			
Safety measures implemented or specified			
Residual risks defined			
Results of the evaluation Objectives achieved Comparison of risks			

3.8.2 Software tools to support risk assessment

Whilst the above checklist serves as an index to the risk assessment there are benefits to be gained for large projects through the use of a structured database to keep all the information and records in a manageable state.

At least one manufacturer of safety devices has made available a database management package that allows the user to record all the information for a machine risk assessment in software files:

- 'Safexpert' is produced by Sick A.G Safety Systems, Germany. It is a package that carries entry forms and reports to prompt for each stage of the hazard analysis and risk assessment stages of an identified machine project. It includes dropdowns and windows providing access to relevant EN standards. Extensive reporting programs allow the complete dossier of records for the CE qualification history to be printed out or stored.
- Dyadem, a Canadian company, offers a range of hazard study recording tools that incorporate risk matrices. These tools carry extensive dropdown lists of possible hazards that allow users to set up their own scales for a risk matrix. The recorded results show the ranking of each hazard before and after safety measures have been put in place.

3.9 Conclusions

Hazard analysis and risk assessment is the essential foundation to the safety of machines. The procedures are simple but time-consuming and can be tedious. The steps of the process must be recorded to show that it has been done in good faith and to the best of ability with the experience available.

The procedures are made much easier where a machine is of a well-established type since there will generally exist a type C standard that will incorporate risk reduction measures for the most common hazards of that type of machine. Using a type C product standard does not remove the obligation to perform a risk assessment but it should make the task a lot easier.

Risk assessment will be of greatest value to project teams building up a complex machine of production line where a novel set of operating conditions and possible hazards is likely to arise.

One of the cardinal principles in this exercise is …

> Manage risk, reduce surprises

i.e., improved reliability and availability (decrease in downtime due to forced maintenance, accidents, etc.) and thus increased productivity for the end user.

References

[1] Standard EN 1050 Safety of machinery – principles for risk assessment 1996.
[2] Standard EN 292-1: Safety of machinery – Basic concepts, general principles for design – Part 1: Basic terminology, methodology.
[3] Standard EN 292-2: Safety of machinery – Basic concepts, general principles for design – Part 2: Technical principles and specifications.
[4] Sick: Safe Machinery Practical Guide, pp. 9–17.
[5] PILZ: Guide to Machinery Safety, pp. 5–72 .
[6] Guardmaster Safety Navigator: interactive machinery safety reference source, Ver 2.0.

4

Design procedures for safety controls

Contents summary

- Introduction to design standard EN 954
- General strategy for the design of safety controls based on EN 954
- Failure modes of safety controls
- Reliability and availability principles
- Measures for risk reduction
- Steps in the selection and design of safety measures based on EN 954
- Explanation of safety categories and SILs (Clause 6 of EN 954)
- Specification of safety requirements and selection of categories
- Examples of circuit arrangements for safety categories
- Practical exercise. Evaluate examples.

4.1 Introduction to design techniques

In this chapter we jump straight into the business of designing safety-related controls for machines. This activity is central to our workshop because most of the workshop participants are likely to be working in electrical or control systems design or in maintenance roles. If you are in maintenance position it is equally important to understand the design principles behind the equipment you are asked to maintain. We need to establish the theory of safety systems design before we move on looking at the various devices available to use for practical applications.

4.1.1 Purpose of this module

The purpose of this module is to learn how to convert the initial requirement for a safety control function into a practical design for an SRECS that will be compliant with the EN standards. Along the way we would like to gain an understanding of the factors affecting the conceptual design of a safety-related control system.

In particular, our learning objectives here are:

- Know about a systematic design method to specify SRECs
- Become familiar with key safety function characteristics and their basic features

- Be aware that in safety controls we as much concerned about how they fail as we are about how they work
- Understand safety categories and their role in safety controls
- Be able to analyze and design circuit arrangements suitable for various categories
- Know how to select the right category of device for an application.

4.1.2 Introducing EN 954-1

As soon as a safety function has been identified, we have to decide if it is feasible to build a practical solution. This usually requires that we sketch out one or more possible designs for the safety controls and then evaluate them for factors such as reliability, failure modes, maintainability and cost.

If this seems a bit complicated the good news is that there is a well-established design standard that offers a reasonably simple and systematic approach to the design of safety-related parts of control systems. This is the Harmonized European Standard EN 954.

In this chapter we make extensive reference to EN 954 and we work through its design procedures. We shall then look at the safety categories defined in the standard for the safety-related parts of electrical control systems. The categories give us a simple scale of safety performance for subsystems and have proved to be very popular with manufacturers of safety system components because they can offer products with a recognized grading in performance.

4.1.3 Design objective

Our objective is to be able to ensure that the design of the safety system satisfies all requirements for an appropriate level of safety performance whilst being a trouble-free and cost-effective solution. At the same time the end result has to be seen to have come from a logical and documented design process typically as specified in EN 954.

4.1.4 Position on programmable systems

We need to be cautious as we proceed with using the EN 954 standard as it has become generally accepted that this standard in its present form does not meet the need to deal more specifically with complex safety systems involving programmable systems such as PLCs. We shall investigate this issue in Chapter 7. EN 954 in its present form represents the best-established principles for safety-related controls in machinery applications where the controls are based on hardwired equipment. It is at present being overhauled to bring it into line with the newer standards for programmable safety systems and this means that is going to have a long life in the industry.

4.2 Review of design standard EN 954-1

This standard was published in 1997 and is intended to provide guidance during the design and assessment of control systems. It is listed as a type B1 standard which means that it follows the design principles of the type A standards and can be used to give guidance to technical committees preparing application-specific standards such as the type C standards we have seen in module 2.

This is very useful to us because if we understand the principles used in this standard, we can more effectively use a type C standard where one of them is applicable to a particular machine. It also means that if we follow this standard in our design process for a new application we shall be seen to be using a widely accepted code of practice.

4.2.1   Scope of EN 954-1

The scope of EN 954-1 is as follows:

- Provides safety requirements and guidance on the principles for the design of safety-related parts of control systems
- It applies to all safety-related parts of control systems, regardless of the type of energy used, e.g. electrical, hydraulic, pneumatic, mechanical
- It applies to all machinery applications for professional and non-professional use
- Provides an iterative design procedure and cross-references key standards
- Defines five categories of safety performance for control systems and subsystems
- Requires validation procedures to be carried out and specifies essential tests.

The standard is best known for its description of the five safety categories for control systems. The categories are based on the ability of the controls to perform under fault conditions. This will clearly be related to the degree of risk reduction that can be expected from the controls, i.e. the greater the risk reduction we need, the higher the safety category we need.

4.2.2   EN 954 useful annexes

Annex A is a checklist enunciating some basic requirements to be adhered to during the process of selection and design of safety measures. The guidelines cover:

- Assumption of type of fault location in components
- Selection of correct reference category of fault
- Actions to be initiated by SREC on occurrence of any type of fault – systematic and random
- Adequate means of meeting maintenance requirements
- Reduction in risks
- Refinement in ergonomic of safety measures (devices) and other requirements like reliability, availability, etc., such that these are easily maintained during life cycle of machinery.

Annex B provides substantial guidance and a chart to assist selection of safety categories for the control system parts. We shall detail this in a moment but before we get to the category stage let's go through the design steps given in the standard and take a look at the issues that are raised. Later we can repeat the exercise with an example problem to see how it works out.

Annex C lists some of the salient faults and failures for various technologies like:

- Electrical and electronic components
- Hydraulic and pneumatic components
- Mechanical components.

Annex D elaborates relationship between safety, reliability and availability for machinery.

4.3   Procedure for the design of safety controls based on EN 954

Figure 4.1 below is a simplified version of the one given in the standard. It shows the five key steps involved in arriving at a safety systems design.

The design procedures it describes are based on the systematic approach defined in EN 1050 leading to the identification of individual safety functions. For each safety function

EN 954 then describes how each safety function is to be developed into particular functional requirements and how each function must be given a safety category. The five-step design cycle is summarized below. For a more detailed understanding of the steps, we recommend readers to examine EN 954-1, Clause 4.

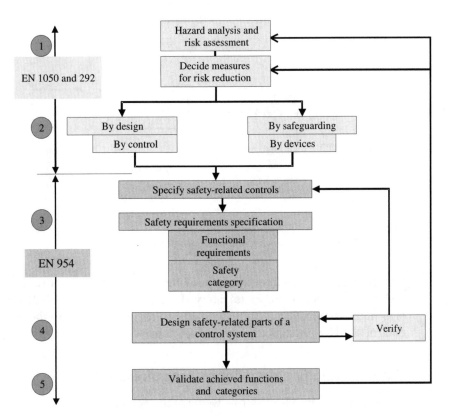

Figure 4.1
The iterative design process for safety-related parts of control systems

4.3.1 Step 1: Hazard analysis and risk assessment

The hazard analysis and risk reduction is the first step in design process to take stock of the dangers likely to be encountered when the machine is in use or being maintained. We covered the procedures for doing these studies in Chapter 3. This step should produce a list of hazards with their risk estimates, typically in the form of a risk-ranking table together with a decision on the need for risk reduction.

4.3.2 Step 2: Decide measures for risk reduction by control means

This step is as per EN 292-1 as described in our module 3 and identifies the risk reduction measure as being in the design and by safeguarding. Those measures provided by control means are the one we are interested in here. The result of this step is the list of safety functions each identified by a simple description.

Example 1 E-stop requirement: Operator must be able to stop the conveyor from any position along its length using one hand whilst standing within 500 mm of the rolls.

Example 2 The cutting wheel must be prevented from starting and running if the guard is not fully closed. The guard must be locked closed automatically whilst the cutting wheel is turning. The cutters have a high inertia and may spin for several seconds depending on load. Therefore the guard must remain locked in place until motion has stopped. This is to be achieved by a motion-sensing device.

It is important to note here that any control system function that makes a significant contribution to safety must be regarded as '*safety-related*' (*see Figure 4.2*). *If the function is already part of the machine control system it must still be identified for treatment as safety-related.*

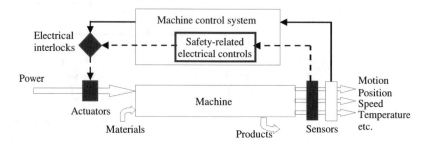

Figure 4.2
All control functions that provide safety are identified as safety-related

Once a control function is safety-related it becomes subject to the requirements of EN 954 or any other appropriate safety design standards. We shall see later that this has significant implications for PLC-based control systems due to the requirement that safety control shall be functionally separate from basic controls.

4.3.3   Step 3: Specify safety control requirements

Translating this into basic terms we have to decide:

- What sort of safety function it is? … See para below on characteristics of safety functions.
- How it is going to be arranged? This results in a block diagram of the safety control design as we shall see in a moment.
- What safety category it must have … Define the safety performance of the controls.

4.3.4   Step 4: Design the safety controls

Here we can proceed with the selection of the equipment and the circuit arrangements for it to function as specified. The design can proceed using the specific requirements laid down in standards (e.g. for an E-stop device: refer to EN 292, Annex A, 1.2.4 and Para 9.2.2 of EN 60204-1) and selecting a device or subsystem to meet the category defined in Step 3.

4.3.5   Step 5: Validation

Note that validation is a fundamental principle of all safety system design processes. It is the disciplined practice of formally checking that the end result of the design process still matches up correctly to the original requirements as defined at each step. This has to be achieved by analysis of the design, and then by analysis and by functional testing of the safety system for all normal and foreseeable abnormal conditions.

Validation activities are detailed in paragraph 8 of EN 954.

- *Prepare validation plan:* This describes the analysis of the design for failure modes and its ability to meet the specified safety category. It should then cover the validation by testing including testing of the functions, proving the categories, proving the dimensions or parameters of the safety function and its compliance with environmental parameters.
- *Validation by analysis:* This refers to proving that the reduction in risk can be achieved. This involves running through the fault lists to include reasonable failure modes and excluding those that are improbable. The analysis methods can use fault tree analysis, failure mode and effect analysis (FMEA) and FMECA (criticality analysis) and a checklist of systematic faults (see Chapter 9).
- *Validation by testing:* This is functional testing of the installed safety system for all normal and foreseeable abnormal conditions.
- *Compliance testing:* This demonstrates that the equipment will perform in the presence of expected environmental conditions. These will typically include vibration, shock loading, EMC (interference) and the effects of materials being processed.
- *Validation report:* The report is an essential component of validation and it will record details of the equipment and all the validation activities with results. This report will be used in the technical construction file used for the CE compliance processes and it is important that it will be written in a form that will be auditable by other parties.

4.3.6 Outcome of the design procedure

The above five-step design procedure, if properly executed, should result in a safety control system that can be shown to provide a satisfactory degree of risk reduction matched to the original risk assessment. The documented records of the design will provide a traceable history showing the reasons for each feature and recording the factors that were considered by the designers. The records will show where assumptions have been made or where some fault conditions were excluded.

4.4 Design considerations

Here we take look at some of the key points raised in EN 954 concerning the basic design of the safety controls. These points are useful because they apply to most safety application projects.

4.4.1 Safety objectives (Refer 954, Para 4.1)

The safety-related parts of a control system are to be designed and constructed using the principles of EN 1050 and should allow for all circumstances such as:

- Their intended use and foreseeable misuse
- When fault occurs in the safety system equipment
- When foreseeable human mistakes are made during the intended use of the machine as a whole.

4.4.2   General design strategy (Refer 954, Para 4.2)

- Use risk reduction measures in hierarchy as defined by EN 292-1.
- The category of safety controls will depend on the contribution made to risk reduction.
- Explains that greater risk reduction requires higher resistance to faults being built into the design of the controls.
- Suggests that redundant structures may be better than high-reliability single-channel designs.

For example, compare the failure rate of a single-channel switch with dual-channel redundant switch architecture.

Figure 4.3 shows a single limit switch as an input to a safety system. The safety guard door must be closed to allow the contacts to close and hence allow the machine to be started. When the door is opened exposing the hazard, the circuit is opened and the machine must stop. The fail to danger mode is a short-circuited switch or burned through transistor.

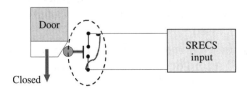

Figure 4.3
Single-channel guard door limit switch with fault; safety function fails

Suppose MTBF = 11 years, i.e. fail to danger rate = 0.09 faults per year or 1.0E − 05 faults per hour.

If the switch and its circuit is tested once per month and if it is repaired within 8 h the probability of a dangerous failure per hour is approximately 5.0E − 06 or 1 failure per 200 000 h. (Reliability calculations based on formulae in IEC 61508, Part 6, Table B2.) This translates to one failure per 22 years.

Figure 4.4 shows a pair of the same switches arranged in parallel input to a safety system. For the same fail to danger rate for each switch the dangerous failure condition will only occur if both switches develop the same type of fault in the same month between testing.

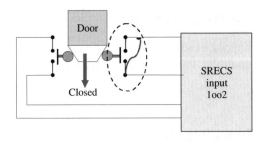

Figure 4.4
Dual limit switches shown with single fault; safety function still works

The dangerous failure rate per hour is calculated to be approximately: 5.2E − 07 or 1 failure per 2 000 000 h. This translates to one failure per 220 years.

If we assume that 220 years between faults is an acceptable fault rate you can see that the choice of slide 2 may be much better than trying to find a single switch with a very high reliability. EN 954 is suggesting that you might use a less reliable switch in a redundant arrangement and get a better result.

4.4.3 The problem of common mode failures

We shall see later when we touch briefly on reliability analysis that providing redundancy not always a guarantee of improved reliability. For example, the above calculation has allowed for the fact that 10% of the switch faults may occur as common faults in both switches. In some cases the common factor may be much higher than that. It's also possible that the testing method may be imperfect, adding to the chances of failure of both switches. In Figure 4.5 a single limit switch has been used with two contact poles wired in series. The common mode fault is a stuck plunger. This design has some redundancy in the electrical circuit but both circuits suffer from the common mode failure of the plunger and hence will not be much better than a single-channel design. Obviously two independent switches will do a much better job here.

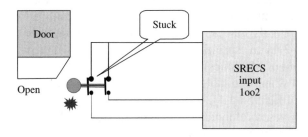

Figure 4.5
Two-pole limit switch with common mode fault: safety function fails

4.4.4 Failure modes of safety controls

It may be useful at this point to recall the typical faults to be considered when deciding on the structure of the control system and its devices. We refer here to the guide to faults listed in Annex C of EN 954.

(a) Electrical/electronic components:

- Short circuit or open circuit in cables or conductors
- Short circuit or open circuit occurring in single components
- Non-drop-out or non-pick-up of electromagnetic elements
- Mechanical blocking of moving elements, loosening or displacing of fixed elements
- Drift beyond the tolerance values for analog elements
- Loss of entire function or part of function in complex integrated components.

(b) Hydraulic and pneumatic components:

- No switching or incomplete switching of the moving element
- Drift in the original control position of moving element
- Leakage and modification of the leakage volume flow
- Unstable control characteristics in servo-valves and proportional valves
- Loss of pressure or bursting of lines
- Clogging of the filter element

- Abnormal pressure and/or volume flow
- Failure or abnormal modification of the input or output signal characteristics in sensors.

(c) Mechanical components:

- Loosening of fixtures
- Sticking of guide moving components
- Misalignment of parts
- Spring.

4.4.5   Characteristics of safety functions

This term refers to the essential nature of each safety function, e.g. an E-stop or a remote control for starting. Table 1 of EN 954 contains an extensive cross-reference to where essential characteristics are defined in this and in other standards. This helps us to know where the ground rules for each characteristic can be found. For example, inspection of Table 1 reveals that nearly all electrical functions are defined in EN 954 itself and in EN 292-2 with specific detail in EN 60204-1 (Safety of Machinery – Electrical Equipment of Machines).

Here is simplified table showing the characteristics and some of references provided by Table 1 of EN 954-1.

Note: Y= yes, a reference is made in this standard. Blank = no reference. A = Annex A in EN 292-2 which is also Annex I to the EC Machinery Directive relating to EHSR for machinery and safety components.

It is clear that Table 1 in EN 954 provides a valuable reference tool. It also indicates that engineers and technicians working in the machinery safety field should have, as a minimum, a set of standards that include EN 1050, EN 292-1, EN 292-2, EN 954 and EN 60204-1. In particular paragraph 9 of EN 60204 -1 "Control circuits and control functions" is a very useful and definitive reference for the essential characteristics of most of the commonly used control functions.

Safety Function Characteristics	EN 954	EN 292-2	EN 60204-1	Other
Design principles	Y	Y A	Y	EN 60335
Ergonomic principles	Y	Y A	Y	
Stop function	Y	Y A	Y	
E-stop	Y	Y A	Y	
Manual reset	Y	A	Y	
Start and reset	Y	Y	Y	
Response time	Y			EN 999
Safety-related parameters	Y	Y	Y	
Local control function	Y	Y		
Muting	Y		Y	

Safety Function Characteristics	EN 954	EN 292-2	EN 60204-1	Other
Manual suspension of safety functions	Y	Y A	Y	
Fluctuations, loss and restoration of power source	Y	Y A	Y	
Programmable electronic systems		Y	Y	
Unexpected start-up		Y A	Y	
Indications and alarms		Y A	Y	
Escape and rescue of trapped persons		Y A		
Electrical equipment		A	Y	
Electrical supply		A		
Other supply		A	Y	
Covers and enclosures			Y	
Pneumatic and hydraulic equipment		Y A	Y	EN 982-3
Isolation and energy dissipation		Y A	Y	EN 1037
Physical environment and operating conditions		Y	Y	
Control modes and control selection		Y A	Y	
Interfaces/connections			Y	
Interaction between different safety-related parts of control systems		Y	Y	
Man–machine interface		Y A	Y	EN 60447

4.4.6   Key points for basic control functions

Part 5.2 of EN 954 includes some useful basic requirements for common safety functions. Some of these are listed here to remind us of the basic principles to be observed when calling up or checking such controls.

Stop function initiated by a protective device

These are the commonly used type of stop functions seen when a machine stops if a guard is opened or if a light curtain is interrupted. These stops may be controlled stops allowing simple restarts or they may operate in the same way as E-stops. We will see more on this in Chapter 5.

The stop function after being actuated will put the machine in a safe mode such that danger or hazard to human is eliminated. This stop function will enjoy priority over the normal operating stop function. In the case of a group of machines working together the safety logic may need to take suitable measures to keep the overall group in a safe state. Stop functions such as these are very important in automation systems since an un-coordinated stop of a set of linked machines or a sudden stop of a complex machine can be damaging.

E-stop function

The difference between a stop function and an E-stop is that the E-stop is a manually operated stop and it has priority over any other form of controlled stop. EN 954 also requires that this action does not lead to hazard at the interface points with other machines. If this happens the E-stop function will have to be linked to the othermachines.

The E-stop condition must also have the facility to be signaled as status to other linked machines.

Manual reset

The protective devices which are required to be of manual reset type shall on receipt of the protection trip command continue to remain in operated state (stop condition) till the manual reset device is reset and safe conditions for starting the machinery have been restored. The risk assessment shall indicate the use of a device having manual reset. The manual reset function:

- Shall be provided through a separate and manually operated device within the safety-related parts of the controls
- Shall only be achieved if all safety functions and protective devices are operative. If this is not possible the reset shall not be achieved
- Shall not initiate motion or a hazardous situation by itself
- Shall be by deliberate action
- Shall prepare the control system for accepting a separate start command
- Shall only be accepted by actuation of the actuator from its released (off) position.

It is of paramount importance that this reset actuator is installed in safe zone from where all of the machinery is visible so that operator can check for the danger zone being free from persons.

Start and restart

Rules for starting of machines are referenced to EN 60204-1. Start and restart rules also apply to remotely controlled machines. Essentially the start of a machine shall only be possible when all safeguards are in place and are functional. This important point highlights the need for safeguards to have means of proving that they are healthy before a start is permitted. This is where self-diagnostic circuit checks begin to play an important role in the design of safety circuits.

Response time

Response time requirements are referred to EN 999. Response times apply to the time between detection of a hazard and the moment when the machine is brought to a safe condition. The response time of the protection device control system has to be furnished by designer or supplier if risk assessment requires it to be declared. This shall be in addition to overall response of machine.

Safety-related parameters

Safety-related parameters are the settings used in safety system to define the limits of physical conditions such as travel, speed or pressure. These must be carefully set up in the control system. In some applications there will be need for the limiting values to be changed according to the process or the product being made. The standard refers to requirements defined in EN 292-2, Para 3.7.9. These require that a control mode selection switch be used to change settings and that the switch should be lockable in the set positions.

If errors in manual inputting of safety-related data in programmable electronic systems can lead to a hazardous situations, then a data checking system within the safety-related control shall be provided, e.g. check inputs, format and/or logic input values, etc.

When safety-related parameters, e.g. position, speed, temperature, pressure, deviate from preset values/inputs, the control system shall initiate appropriate measures, e.g. functions of stopping, warning, alarm, etc.

Local control function

When a machine is controlled locally, e.g. by a portable control device, pendant, the following requirements shall also apply:

- The means for selecting local control shall be situated outside the danger zone.
- It shall not be possible to initiate hazardous conditions from outside the zone of local control.
- Switching between local and external, e.g. remote, control shall not create a hazardous situation.

Muting

Muting is the term applied when a safety function is temporarily suspended or modified to allow an action that will not be hazardous at that stage of the machine's operation. For example, if a light screen protects an operator from placing his hands in the tool area of a press it may be switched out once the press has closed or has started the up stroke. Muting is a major feature in light curtain applications as it contributes to high productivity.

The selection of parts or components for the function of muting should be such that they do not degrade the safety integrity of the safety function, i.e. the muting circuits are just as critical for safety as any part of the safety system.

It shall be ensured that no person is exposed to dangers or hazardous situations whenever muting conditions are actuated for any safety-related device. During this operation safe conditions shall be ensured by other means. This may sometime necessitate giving an alarm or warning whenever muting is initiated. Once the muting condition has been removed or de-actuated all safety functions of the safety-related parts of the control system shall be restored.

Manual suspension of safety functions

During testing, machine set-up, maintenance, troubleshooting, etc., it may be necessary to manually suspend safety functions. The requirements as stated hereunder shall be ensured:

- Effective and secure means to prevent manual suspension in those operation modes where it is not allowed
- Selection of the safety-related parts of the control system which are responsible for the manual suspension such that the principles of EN 1050 are fully taken into account.

In some situations such suspensions need to be initiate either alarm or warning based on risk assessment studies.

Fluctuations, loss and restoration of power sources

During fluctuations in input power or energy levels (even loss) to parameters outside the design values the standard requires that the safety-related parts of the control system should still ensure safe conditions. This means that the safety system will have to be of a fail-safe design in the event of power loss or power dips and surges.

4.5   Safety categories

You will be aware that manufacturers of devices for machinery safety duties often state the safety category rating of each product. This means that the device has the right characteristics to make it suitable for use in a safety-related control function that is to be built to a category value equal to or lower than the one quoted for the device.

The safety category is a way of defining the required behavior of safety-related parts of a control system in respect of its ability to resist faults. The ability to resist faults increases the likelihood that the safety function can be relied upon to do its job and hence provide risk reduction. Roughly, the more risk reduction you need, the higher the safety category for your design.

Safety category is related to safety integrity and to SIL as noted earlier. For the moment we shall concentrate on safety categories as these form the basis of most of the existing safety systems in service except where programmable system are being used.

Paragraph 6 of EN 954-1 defines the characteristics of five safety categories in terms of their component qualities and their degrees of redundancy, self-checking and fault tolerance.

4.5.1   Summary of each category

As explained in the table below these can be divided into five categories, but it is important to note that these categories are not risk levels, nor are they intended to be used in any given hierarchy:

- *Category B:*   A single fault can lead to loss of safety function. The selection of components should be such that they can withstand all the effects of operating conditions. Hence use of standards and test data is a good check for conformity.
- *Category 1:*   Similar to category B but distinguished by the use of well-tried components and well-tried safety principles. For categories B and 1, it is fair to say that the safety of the system relies on the components and their design.
- Category 2:   Employs components that satisfy category B and category 1 but safety performance is also supported by techniques of checking the safety function at start-up and sometimes periodically during use. Category 2 devices are single-channel, non-redundant so that a single undetected fault can cause the loss of the safety function.
- Category 3:   Safety performance is supported by configurations that are single fault tolerant. This means that the device or system will still perform its safety function in the presence of an undetected dangerous fault. In practice this requires dual redundant parts and circuits, as we shall show later. The

equipment is designed to automatically detect some but not all of its possible faults at the time of switching on the safety system and hence there remains the possibility that an accumulation of faults (i.e. two or more) will lead to the loss of the safety function.

- Category 4: The performance is single fault tolerant as category 3 but the fault detection should detect all possible faults or the design should ensure that an accumulation of faults does not lead to loss of the safety function. Reasonable limits are set on the number of faults this rule applies to and it is left to the designer to decide depending on the nature of the equipment.

4.5.2   Table of categories

The table as given below indicates a brief summary of the requirements for the various categories. The complete text for the requirements is contained in EN 954-1, 'Safety-related parts of control systems' Section 6, 'Categories':

Category	Concise Statement of the Requirements	System Behavior	Principle to Achieve Safety
Category B (see note 1)	Safety-related parts of machine control and/or their protective equipment, as well as their components, shall be designed, constructed, selected, assembled and combined in accordance with relevant standards so hat they can withstand the expected influence	When a fault occurs it can lead to a loss of the safety function	By selection of compo-nents (toward prevention of faults)
Category 1	The requirements of category B apply together with the use of well-tried safety components and safety principles	As described for category B but with higher safety-related reliability of the safety-related function (the higher the reliability, the less the likelihood of a fault)	
Category 2	The requirements of category B and the use of well-tried safety principles apply The safety function(s) shall be checked at machine start-up and periodically by the machine control system. If a fault is detected a safe state shall be initiated or if this is not possible a warning shall be given	The loss of safety function is detected by the check The occurrence of a fault can lead to the loss of safety function between the checks	By detection of faults

Category	Concise Statement of the Requirements	System Behavior	Principle to Achieve Safety
Category 3 (see notes 2 and 3)	The requirements of category B and the use of well-tried safety principles apply The system shall be designed so that a single fault in any of its parts does not lead to the loss of safety function	When the single fault occurs the safety function is always performed Some but not all faults will be detected An accumulation of undetected faults can lead to the loss of safety function	By redundant structure and detection of faults
Category 4 (see notes 2 and 3)	The requirements of category B and the use of well-tried safety principles apply The system shall be designed so that a single fault in any of its parts does not lead to the loss of safety function The single fault is detected at or before the next demand on the safety function. If this detection is not possible then an accumulation of faults shall not lead to a loss of safety function	When the faults occur the safety function is always performed The faults will be detected in time to prevent the loss of safety functions	

Note 1   Category B in itself has no special measures for safety but it forms the base for the other categories.

Note 2   Multiple faults caused by a common cause or as inevitable consequences of the first fault shall be counted as a single fault.

Note 3   The fault review may be limited to two faults in combination if it can be justified but complex circuits (e.g. microprocessor circuits) may require more faults in combination to be considered.

4.5.3   Block diagram models of the categories

It may be helpful to look at the practical implications of the safety categories in terms of block diagrams for a generalized safety function. Here is a generalized model we can use for visualizing a typical safety function. Note that the SREC function comprises all stages of the loop from sensor to final control of the drive.

Here we present the configurations for categories 1–4 represented as a complete system of sensors, logic solvers (relays or safety PLC) and actuators (see Figure 4.6). In this case a guard door limit switch must be closed to allow the final drive of the machine to operate.

This category 1 safety function uses a single channel at each of the stages of the safety system (see Figure 4.7). It will use components that are well-tried working on proven fail-safe principles for the circuits and employing positive action devices that avoid any indirect linkages between moving parts. The correct functioning is tested periodically by a manual test, e.g. open the door and check that the drive stops.

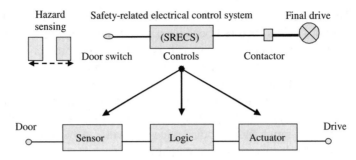

Figure 4.6
Safety-related control system parts resolved in general model

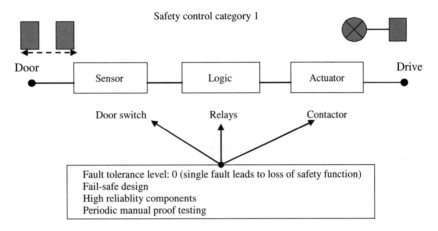

Figure 4.7
Category 1

The category 2 example is also a single-channel system but the logic stage has been arranged such that it will not start-up into an enabling state if any of the input or output circuits are not in their expected states at the beginning of the operation (see Figure 4.8). Usually this indicated by an alarm lamp or the test is initiated when the start button is pressed.

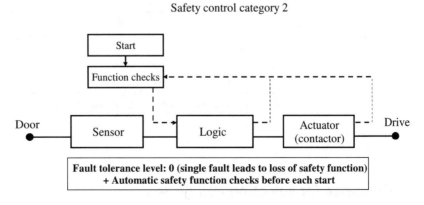

Figure 4.8
Category 2

The category 3 system features redundant devices at each stage of the system (see Figure 4.9). This required because a single fault anywhere in the scheme must not disable the safety function. If contact 1 is shorted the protection remains through contact 2 opening to trip the machine. If contactor (actuator) 1 is stuck in the drive will still trip because contactor 2 will drop out on command. The logic may simply be one or more relays in each channel or it may be a safety certified PLC channel. Fault detection is achieved by self-checking at the start-up and by comparison of the status of the parallel channels of input and output devices as well as the logic. Any discrepancy represents a fault and leads to a trip of the machine.

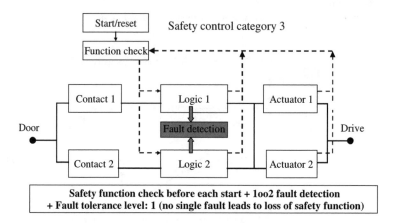

Figure 4.9
Category 3

The category 4 system shown here is similar to category 3 but the fault detection capabilities have been increased to cover all practicable faults and the diagnostic tests are performed at frequent intervals during operation (see Figure 4.10).

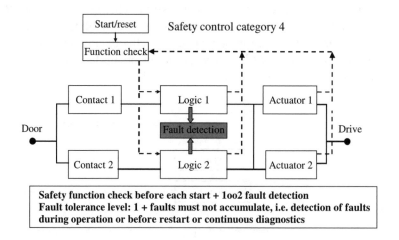

Figure 4.10
Category 4

4.5.4   Guidance for selecting categories

So how do you decide on which category you need? In machine safety the approach, until now, has been to use qualitative estimating of risks rather than quantitative methods. This is probably due to the complex nature of the risks in terms of types of hazard, severity of harm and exposure of persons. The elements of risk (as set out in EN 1050) and the route to risk reduction can be considered as shown in Figure 4.11.

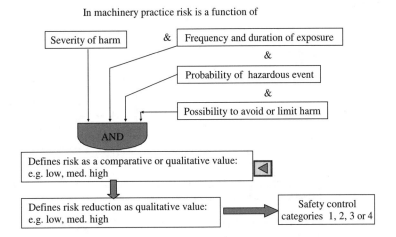

Figure 4.11
Evaluation and reduction of risk in machinery practice

The designer of a machine or automation plant with an assembly of machines is required to perform the risk assessment and arrive at a documented conclusion on the risk factors. The amount of risk reduction required from all the safety measures including the control systems must then be identified and described. Note that this procedure is mandatory for compliance with machinery regulations in EU and is commonly applied in many industrialized countries.

No quantitative scales of measurement have been available either for the risk estimate or for the 'safety reliability' of the controls needed to reduce the risk. The machine builders must take the responsibility to show that the machine is 'safe' after applying a suitable set of safety measures if these are needed at all. 'Safe' is a comparative term and appears to come down to showing that the risks are acceptable due to very low probability or very minor level of harm.

For SRECSs EN 954-1 defines a set of safety categories that approximately correspond to the severity of the risk reduction problem which they are intended to manage. The problem is that the risk reduction need is difficult to define and the standard offers a 'blunt instrument' in the form of a decision chart or risk graph to assist with the procedure. The risk graph to be used as a guide to allocating safety categories is shown in Figure 4.12.

The procedure for using this chart is to start at the left and consider first the severity of injury that is likely to result from the hazard being considered. This makes a decision between slight or serious. The condition includes for the long-term effects of the accident.

The frequency and exposure time to the hazard are then decided by a parameter F. Guidance in the standard suggests that if working with the machine involves regular exposure to the hazard the parameter will be F2.

We previously mentioned 'the possibility of avoidance' being a factor in the nature of the hazard. For example: Is there normally some indicator with a chance of response by the operator or is it too fast or obscure for this?

For example, using the category selection chart we could decide that the safety-related controls for our simple interlocking guard should be built to Category 1 based on the decision: S1. However if we feel the injuries could be significant and irreversible we might decide on the parameters: S2, F1, P1, leading to a requirement for Category 2 equipment.

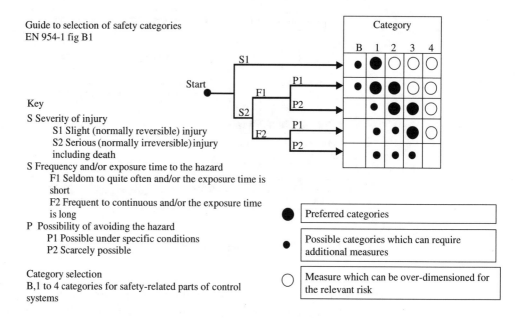

Figure 4.12
Risk graph guide for determination of safety categories as per EN 954-1

The standard emphasizes that this chart is to be seen as an approximate guide only and that each application should be considered on merit and justified through risk assessment. Refer to Annex B of EN 954 for detailed guidance notes on the selection of categories.

We shall see later in Chapter 9 that the new emerging standard for machinery safety using programmable systems is proposing the use of quantitative risk reduction factors as the means of deciding the safety category or safety integrity. This is a developing field and for the present we should work with the basic scheme for categories. EN 954 is to be updated to take into account the programmable systems requirements and readers are advised to watch for developments in the release of an updated EN 954.

4.6   Conclusions

We have outlined the principles of the design life cycle for safety-related control systems based on the five-stage model we saw in Figure 4.1. The process is characterized by careful verification at each step to ensure that the solution being developed is in line with the understanding of the hazards.

In summary:

- Hazard identification leads to risk assessment.
- Risk assessment leads to risk reduction requirements.
- Risk reduction requirements indicate the type and category of safety solution.
- Safety categories indicate the general features and configuration of the safety system.

- Design detail follows the rules of safety categories. Components are available to match-up to the categories.
- Validation establishes that the solution solves the original problem.

In many cases the problem may be quite simple to define and the solution may be equally simple. Many solutions to typical machine safety problems are already well established so that following the life cycle procedures can be fairly simple and it is helpful to refer back to EN 954 as a check on the correctness of the solution. For more original and complex applications, the first principles laid down in this standard can be used with confidence. It is important to remember that EN 954 is a harmonized standard so that for CE marking the designers will want to show that they have complied with the steps described in it.

References

Guidance for material presented in this chapter was sought and is acknowledged from the following sources:

[1] Siemens Safety Integrated Application Manual, Chapter 2.
[2] PILZ: Guide to Machinery Safety, pp. 128–135.
[3] EN 954-1 1996: Safety of machinery – Safety related parts of control systems.
[4] Draft for IEC 62061: Safety of Machinery: Electrotechnical Aspects. Version: 44/380/CD.

5

Emergency-stop monitoring and the safety relay

5.1 Introduction

In this chapter we take a look at the basics of safety relays and the development of monitoring modules that provide what we have called the logic solver function. We shall look into what is given in codes for E-stop so as to understand clearly the requirements imposed on manufacturers, suppliers and users for these safety-related functions.

5.1.1 Objectives

- To know the basic principles of functional stops and E-stops
- To know the basic principles of safety relays
- To know how monitoring relays can help us to meet standards for E-stops
- To understand the essential circuits arrangements for meeting safety categories
- To know how to make use of product selection guides.

5.1.2 Background

Historically, machinery controls used relay-based systems for most of their control functions and safety functions. This still works well for basic single-channel interlocking and sequencing control and still presents no real problems for a simple safety function of category B or category 1. But when we look at the problem of implementing the higher categories in simple safety functions such as the E-stop it becomes a little more complicated.

Monitoring safety relays have evolved to serve this need and are produced in large numbers by several safety control equipment manufacturers. Many variations of safety modules have been developed to support the various safety devices that we shall be looking at in the next module. So at this point it is useful to learn about the basics of the modules.

5.2 Definitions and implications of stop functions

We noted in Chapter 4 that stop functions and E-stop functions are defined in EN 60204-1. Here we need to be aware that there are some important distinctions in terms between stop, E-stop and emergency switching off.

5.2.1 Stop functions

The harmonized standard defines three categories of stop function in paragraph 9.2.2 as follows:

- *Category 0:* Stopping by immediate removal of power to the machine actuators (i.e. an uncontrolled stop – this means that power is removed from all machine actuators, all brakes and/or other mechanical stopping devices being activated).
- *Category 1:* A controlled stop with power to the machine actuators available to achieve the stop and then removal of power when the stop is achieved.
- *Category 2:* A controlled stop with power left available to the machine actuators.

We must note here that the term 'category' has no relation to the term 'safety category' used in EN 954 and described in our Chapter 4. The term we are using here is the 'stop category'. This can be quite confusing when you bear in mind that the stop functions will need to be designed according to the chosen 'safety category'. We could have, for example, a category 2 stop function implemented to safety category 3 requirements.

5.2.2 E-stop

An E-stop is an action to be initiated by a single human operator. The purpose of an E-stop is to protect persons from hazards due to the moving parts or other actions and effects of the machine.

Paragraph 9.2.5.4.2 of EN 60204-1 defines the requirements for E-stop as follows:

9.2.5.4.2 Emergency stop
In addition to the requirements for stop (see 9.2.5.3), the emergency stop function has the following requirements:

- *It shall override all other functions and operations in all modes;*
- *Power to the machine actuators that can cause a hazardous condition(s) shall be removed as quickly as possible without creating other hazards(e.g. by the provision of mechanical means of stopping requiring no external power, by reverse current braking for a category 1 stop);*
- *Reset shall not initiate a restart.*

The emergency stop shall function either as a category 0 stop or as a category 1 stop (See 9.2.2). The choice of the category of the emergency stop shall be determined by the risk assessment of the machine.

Where a category 0 stop is used for the emergency stop function, it shall have only hard-wired electromechanical components. In addition, its operation shall not depend on electronic logic (hardware or software) or on the transmission of commands over a communications network or link.

Where a category 1 stop is used for the emergency stop function, final removal of power to the machine actuators shall be ensured and carried out by means of electromechanical components.

Note that the emergency-stop function must be effective as a category 0 or 1 stop and must take priority over other functions. The stop category must be determined by means of risk assessment. This is a significant feature since it calls for category 0 stops to be totally hard-wired and excluding electronic logic. Where a type C standard calls for a category 0 stop it is not permitted under this standard to use a programmable or electronic device in the design.

Manufacturers of programmable safety systems for machines have argued that a category 1 stop can be just as secure as a category 0 stop and in many cases the controlled stop provided by logic controls is a safer way to stop than a crash stop. Selecting stop category 1 will allow the use of E-stop functions through network connections and PLCs provided all the requirements for safety integrity can be met.

5.2.3   Emergency switching off

The purpose of emergency switching off is to protect persons from the hazards due to electricity. This may be through shock caused by contact with exposed electrical parts or any other hazards and damage caused by electricity.

Paragraph 9.2.5.3.4 of EN 60204 calls for the switching off to be a category 0 stop. This means as above that only an electromechanical device can be used. If this type of stop causes problems the standard suggests that the electrical equipment should have further protection against shock and other effects.

So we can conclude here that E-stops provide protection against functional hazards whilst emergency switching off protects against electrical hazards.

5.3   Safety relay terminology

Safety relay is a term applied to a relay device constructed with 'positively guided contacts' that provides a high degree of confidence in its ability to operate correctly or to fail in a safe manner. The term has become extended to cover modular relay assemblies that provide monitoring and buffering functions between safety sensors and the final element control devices such as the contactors.

5.3.1   What is meant by positively guided contacts?

These are relay contacts that are designed with solid mechanical linkages to provide a fixed relation between normally open contacts and normally closed contacts. This provides confidence that each contact in a set will be in the same position and if one set of contacts is closed the opposite set will be open. An alternative description commonly used is 'force-guided contacts'.

In Figure 5.1, the positively guided contacts are forced to be in one state or the other. Hence it is reasonable to use one pair of contacts as a status indicator for the rest of the contacts. It is also highly improbable that the contacts will fail individually.

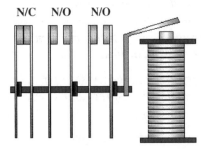

Relay without positively guided contacts

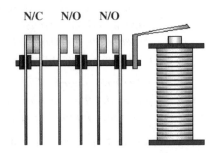

Relay with positively guided contacts

Figure 5.1
Principle of positively guided contacts

The feature of having positively linked or guided contacts ensures that when the relay is arranged with self-checking features there is a high degree of confidence that the status of one pair of contacts will truly reflect the status of the power switching contacts.

The same principle is also applied to E-stop switches and guard switches where the contacts are positively driven by a solid linkage.

The term safety relay is usually extended to cover devices that include two or more relays in a subassembly designed to provide a safety function.

For example, an 'emergency-stop safety relay' is also called an 'emergency-stop monitoring unit'. This is a device designed specifically to provide the safety logic functions needed to perform the E-stop safety functions and provide the self-checking features required by the categories 2, 3, and 4 requirements of EN 954. It is often simply called a safety relay. Often it is called a monitoring relay.

5.3.2   Typical product opdescription

Here's a general description of a safety relay provided in catalog material published by the USA-based company Banner Engineering. This description is helpful for seeing that the requirements for control reliability in the USA match up approximately to those used for category 3 in Europe.

> *The purpose of safety relay model ES-FL-2A is to increase the control reliability of an emergency stop circuit. The ANSI B11.19 standard states:*
> *Control reliability of electrical, electronic, or pneumatic systems frequently consists of multiple, independent parallel or series circuitry or components so arranged that any single failure would not significantly affect the stopping performance of the system or the machine tool.*
>
> *If a failure of a component has occurred, this arrangement either sends a stop command to the machine tool or prevents a successive cycle from being initiated.*
>
> *A typical emergency stop switch offers two redundant switching contacts. In a properly designed machine stop circuit, the opening of either of the two emergency stop contacts immediately removes electrical power from the machine control elements, which react to stop hazardous machine motion and/or any other machine hazard. This redundancy of stopping control offered by a two-pole emergency stop switch is the first step towards control reliability.*
>
> *However, failure or defeat of one of the switch contacts to a short-circuit will go undetected, thereby removing the redundancy, and leaving the emergency stop circuit prone to eventual failure.*
>
> *The ES-FL-2A safety relay connects between the emergency stop switch and the machine emergency stop control elements. The emergency stop switch becomes an input to the safety relay. The safety relay monitors the condition of both contacts of the emergency stop switch. The output of the safety relay consists of two redundant output-switching channels, each of which is the series connection of two safety relay contacts (K1 and K2).*
>
> *The Control Reliability section of ANSI B11.19 goes on to state:*
> *Electromechanical systems that require redundancy and checking of relay contacts should use relays that are designed with mechanical linkages to provide relation between normally open and normally closed contacts to check the contact operation.*
>
> *Safety relays K1 and K2 in the output circuit of the ES-FL-2A safety relay have mechanically-linked contacts which allow the safety relay circuitry to monitor the contacts of K1 and K2 for failure. If the safety relay detects failure of any contact of*

either the input emergency stop switch or the output relays, the safety relay output is disabled and cannot be reset.

The ES-FL-2A safety relay also provides a necessary reset function. ANSI B11 and NFPA 79 standards require that a reset routine be performed after returning the emergency stop switch to its closed-contact position. This prevents the controlled machinery from restarting by simply closing the emergency stop switch.

5.4 How does an E-stop safety relay work?

The basic safety relay arrangement uses an internal latching sequence to set up two or more output relays into an energized condition when all circuits are healthy and after a reset contact has been closed. The relays remain latched in until the input circuit is broken either by the guard door switch or by the E-stop.

A typical hardware-based implementation of a guard door safety function will link the guard door switches in series with an E-stop switch to provide an input to a latching relay. The latching relay will trip when the guard door is opened or when the E-stop is pressed.

To improve the safety of the circuits an additional relay is used to prevent the latching relay from being reset unless the safety control circuits are healthy (i.e. free of dangerous faults). For example, in Figure 5.2 a simplified safety relay is shown where K3 is a relay that must be energized before the latching relay K1 can be set. K3 will not energize unless the power control contactor(s) C has been released, proving that it is not held in by another stray circuit or by a mechanical defect.

In practice relay K1 is usually duplicated by a second channel or redundant relay K2 and both relays must be energized and latched to close the output circuits. K3 is often arranged

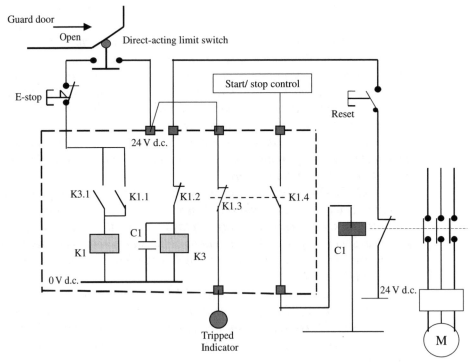

Simplified E-stop and guard switch monitoring relay

Figure 5.2
Simplified monitoring relay

with multiple contacts and expansion units to enable many drives to be interlocked from the same logic.

The example shown in Figure 5.2 uses a safety monitoring relay unit to perform the essential logic functions required to provide safety integrity. These are:

- Checks on the state of input signals. They must be safe before latching reset is allowed.
- Detection of stuck contactors through monitoring of the auxiliary contacts.
- Wiring faults in the input and output circuits.
- Timing and logic for controlling the reset actions, etc.

The safety monitoring relay modules ensure that the safety interlocks and E-stop functions are able to operate independent of the basic control system actions at all times. This is one of the most essential features of any safety control system.

5.5   Practical safety relays

In practice most safety relays incorporate a dual redundant pair of latching relays to improve their fault tolerance level and provide the diagnostic benefits of comparing one relay state with another that should always be the same. In Figure 5.3, we reproduce a diagram of a PILZ PNOZ category 2 safety relay in the form that has been in service for many years.

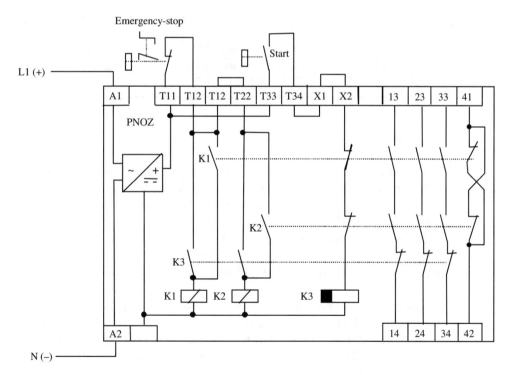

Figure 5.3
A PILZ PNOZ category 2 safety relay

In this basic device the latching function of relay K1 is duplicated in K2. Both K1 and K2 must be properly de-energized, as must K3 for the reset action to work. If one of the relays K1 or K2 is stuck or welded up the device will not permit operations and the output contacts will remain open.

Using this module the user simply hooks up the external circuits in accordance with standard circuit arrangements. Figure 5.4 shows a basic circuit arrangement suitable for category 2.

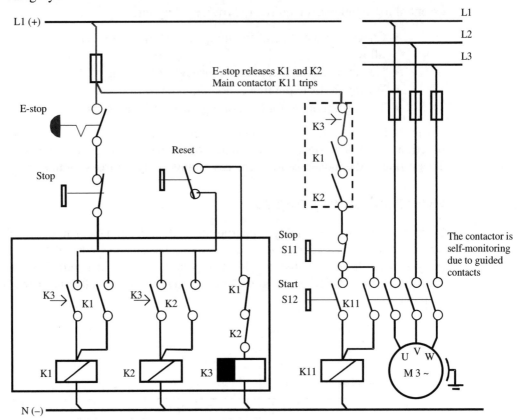

Figure 5.4
A basic safety relay circuit for risk category 2

In normal operation relays K1 and K2 remain latched in whilst the motor can be started and stopped without disturbing the E-stop latched in condition. The requirement for periodic testing is satisfied when the E-stop latch condition is tripped and the module has to be reset before the outputs are enabled.

The basic start and stop controls provide a limited amount of proving of the safety of the final element stage (the contactor K11) because if it has positively guided contacts the latching contact around the start button S12 must be a confirmation of the position of the power contacts.

Figure 5.5 shows how the E-stop monitoring relay can trip several independent motors in a machine. The basic control circuits for starting and stopping each drive are connected in series with contacts of K1 and K2. Multiple sets of relay poles are used to supply one drive per pole. This design serves a machine where all drives must stop when the E-Stop is pressed and where no other safety trips are involved.

5.5.1   Adding additional interlocks

If there is need to provide additional safety functions in a machine it may be necessary to add one or more additional and separate tripping safety relays to selectively stop one or more drives. This is simply achieved by connecting the tripping relays in series as per the example shown in Figure 5.6.

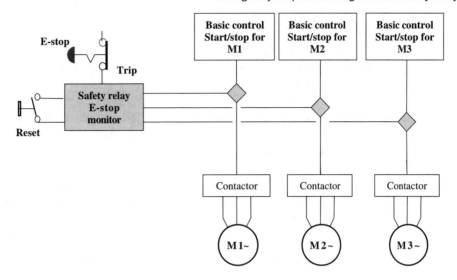

Figure 5.5
Safety relay: E-stop function common to drives, category 2 (without monitoring of final control)

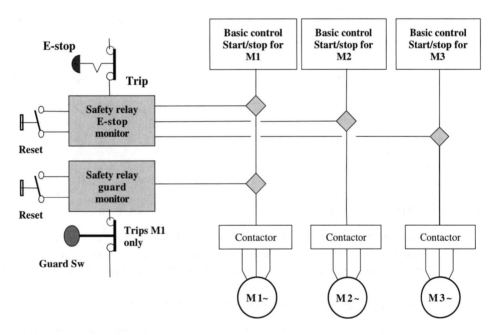

Figure 5.6
Adding a selective trip through a guard-monitoring relay

In the figure shown here there is no option to share the guard switch and E-stop modules, as the logic requires that only one drive be stopped when the guard is opened. In some applications the guard monitor relay may be required to be self-resetting. A further advantage of having separate monitoring relays for each function is that the functions can be identified easily and status indicators can be linked in to assist with rapid fault finding.

5.5.2   Monitoring of final elements

Safety functions built to risk category 2 require that faults should be detected by a periodic safety function check. The safety function check is improved if the status of the final control element is confirmed before the monitoring relay is latched in by the action of pressing the reset button. Hence it is good practice wherever possible to include in the reset loop feedback contacts from the final contactors. Reset of the E-stop system is not possible if either of the contactors is still energized or is jammed on. This creates a high level of fault detection in the output stage of the safety function.

Feedback monitoring is shown in Figure 5.7 for a category 2 application with three drives sharing a common E-stop function. Category 3 requires that faults should be detected before the start of next machine cycle. This requirement makes it essential check the correct status of the final elements by feedback.

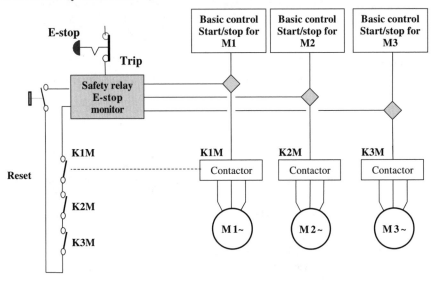

Figure 5.7
E-stop function common to drives, category 2 (with feedback monitoring of contactors)

5.5.3   Monitoring of valves

Similar monitoring functions can be achieved with other final elements such as hydraulic or pneumatic switching valves. In these cases a limit switch on each valve confirming that it is in the safe position before reset can be performed will achieve the monitoring.

5.5.4   Category 3 E-stop monitoring safety relay

In accordance with the principles of categories 3 and 4 safety subsystems the monitoring safety relay must provide for the E-stop function to be fault-tolerant. In practice this requires at least dual redundancy at each of the three stages of the safety function as we saw in Chapter 4 (sensor, logic and final actuator stages). This is achieved in an E-stop function by having:

- Two independent sets of input contacts from the E-stop switch
- Two independent relays in the safety module
- Two independent contactors or valves in the final element stage.

Figure 5.8 depicts the arrangement of the safety relay having two inputs from the E-stop switch and operating two contactors in series with the drive. The reset loop includes feedback monitoring of the contactor states as described in the preceding section.

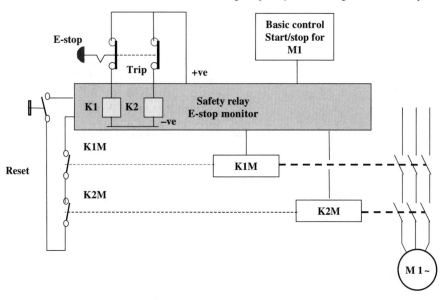

Figure 5.8
Category 3 safety relay's external connections

The dual E-stop contacts operate identical redundant safety relays inside the monitoring module (not shown in diagram). This arrangement is suitable for category 3 applications because it provides a means of detecting a single dangerous fault in the input circuit and it provides self-test each time the circuit is 'armed'.

The dangerous faults could be a short across either of the input contacts. However it does not protect against wiring faults where two contacts are shorted between the two sets of contacts in an E-stop switch. This is shown in Figure 5.9.

If this occurs and subsequently there is a short across one of the contact pairs, the protection will fail. This is an example of an 'accumulation of faults' as defined in EN 954 and hence the device rating is limited to category 3.

The type of fault shown here will not be detected because the input relays in the safety monitor are arranged in 'identical redundancy'. They will both respond identically to the fault and will continue to operate normally (see Figure 5.10).

Monitoring relay: category 3 detection of faults

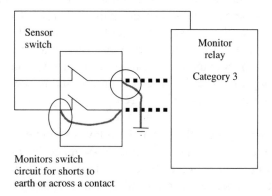

Figure 5.9
The detection of input wiring faults of a category 3 monitor relay

Monitoring relay: category 3 detection of faults

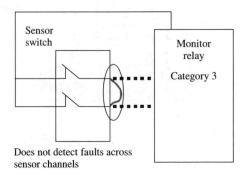

Figure 5.10
Wiring fault undetected by a category 3 monitor relay

This design will continue to operate without detecting this fault so that if there is another defect such as a short across the switch contacts the safety function will fail. This is described in the standards as an 'accumulation of faults' (see Figure 5.11).

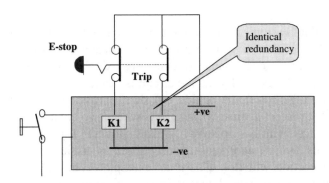

Figure 5.11
Identical redundancy at the input stage of a category 3 safety monitor

5.5.5   Category 4 E-stop monitoring safety relay

Category 4 demands a higher level of fault detection in a dual redundant configuration so that the faults are either detected as soon as they occur or not allowed to accumulate. Category 4 safety relays protect against shorts between the dual channels of the input circuits as well against earth faults and shorted contact pairs. As with the lower categories the start-up test cycle protects against internal faults in the module. The typical arrangements for a category 4 safety relay require the addition of diverse redundancy to avoid the risk of common mode seen in the category 3 designs. Figure 5.12 shows how diversity has been introduced in the circuit arrangements at the input of a category 4 E-stop monitoring relay.

In this arrangement the channel 1 relay is switched via an E-stop contact on the 0 V side of the circuit. The channel 2 relay is switched by the second contact pair of the E-stop switch connected between the 24 V supply and the relay. This arrangement will detect a short between the separate channels of the E-stop switch because the switch contacts are positioned on the positive side of one relay and on the negative side of the other relay. In the workshop we can run through the various effects of faults.

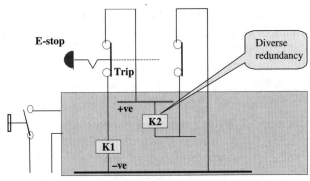

Figure 5.12
Input arrangements for diverse redundancy in a category 4 E-stop monitor

5.5.6 E-stop relay: category 4 (Banner Engineering example)

In Figure 5.13, the same principles are deployed in a category 4 E-stop safety relay from Banner Engineering (USA origin). This unit is very similar in functionality to the PILZ unit and is designed specifically to meet the USA standards for E-stops as defined by ANSI B11.20-1991 and NFPA 79. We shall look at E-stop regulatory requirements in USA and EC later but at this stage it is important to recall that all E-stop functions engineered for the USA market must have the equivalent of category 4 performance to meet the 'control reliability' requirements.

This requirement from the ANSI standards does not allow us the choice of a lower category that is permitted under EU regulations.

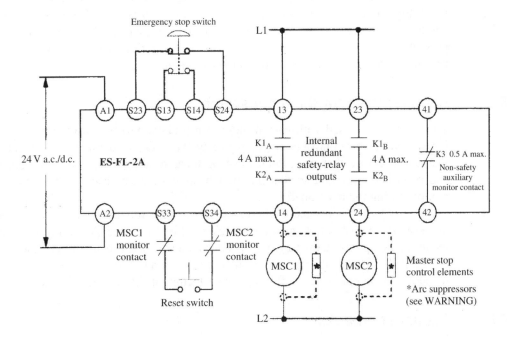

Figure 5.13
A schematic diagram of category 4 E-stop safety relay from Banner Engineering

In this figure, the hook-up shows connections to a dual redundant and electrically separate pair of contacts in an E-stop switch with redundant safety relay outputs to two contactors (final elements).

The following description provided by Banner points out the essential requirement for category 4 applications to continue the fault-tolerant principles through to the actuator or final drive control stage of the safety function.

The hook-up diagram (see above) shows a generic connection of the ES-FL-2A safety relay's two redundant output circuits to master stop control elements MSCI and MSC2. A master stop control element is defined as an electrically-powered device, external to the ES-FL-2A safety relay, which stops the machinery being controlled by immediate removal of electrical power to the machine and (when necessary) by applying braking to dangerous motion (reference ANSI B 11.19, section 5.2. "Stop Control"). This stopping action is accomplished by removing power to the actuator of either master stop control element.

To satisfy the requirements of Safety Category 4 of EN 954-1, the master stop control elements must be safety relays which offer a normally-closed forced-guided monitor contact. One normally-closed monitor contact from each master stop control element is wired in a series connection together with the normally-open RESET switch. In operation, if one of the switching contacts of either master stop control element fails in the shorted condition, the associated monitor contact will remain open. Therefore, it will not be possible to reset the ES-FL-2A safety relay.

5.5.7   Warning note

Banner Engineering points out in a warning note that arc suppressors must not be wired across the output contacts of the safety relay; they must be placed across the coils of the contactors. No device must be added across the output contacts of the module because it could defeat the purpose of both redundant channels at the same time. This is known in reliability modeling as a 'common cause failure' and is one of the most dangerous fault conditions that can arise. We shall see in a later module that common cause failures present limitations on the performance of many redundant systems, hence the strong recommendations for diversity in redundant designs.

5.5.8   Requirements for the reset switch

The reset switch used with E-stop monitoring relays will be a low power device suitable for switching the control voltage, typically 24 V DC and rated for typically 100 mA. Standards require that the switch must be located safely away from the hazardous area and must be positioned so that the switch operator may observe any area of dangerous motion during the reset operation.

The reset switch is normally wired in series with the monitoring contacts or feedback loop in most designs of safety monitoring relay. This raises the possibility of failures of the monitoring loop through wiring or re-connection errors after maintenance. Such errors should be checked as part of the periodic manual proof-testing regime. It is not clear whether some monitoring relays are able to offer separate detection of shorted reset functions.

5.5.9   Expander modules

Suppliers of safety relays offer expander modules to increase the number of safety contact channels available for actuation from the main unit. Each expander module is slaved to the main unit but the essential feature is that it must have a feedback-monitoring contact available for inclusion in the reset loop. This ensures that each separate relay in the set of relays is monitored.

A typical expander module is shown in Figure 5.14. The feedback loop connections are Y1 and Y2, these will be connected in series with the feedback connections on the base unit.

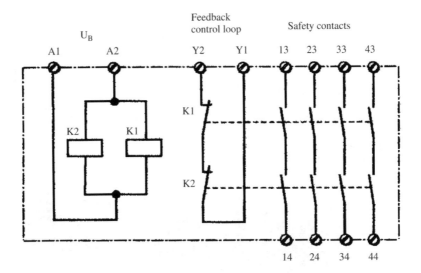

Figure 5.14
A schematic diagram based on a PILZ expander module

5.6   Certification

Safety relay devices are certified as fit for their specified duties by approved organizations including TUV, BG and UL. These laboratories examine and test the devices to verify their compliance with essential requirements as laid down in standards. These include the safety categories as described by EN 954-1. This gives us confidence that the product and the usage instructions that come with it will allow the end user to achieve the desired level of safety integrity.

5.7   Functional overview of monitoring relays

The E-stop relay systems have very similar functions to most of the common safeguarding functions such as the detection of safety guard positions. Figure 5.15 summarizes the general features we have discussed and indicates the typical arrangement of a monitoring relay in a safety-related control system. It has been common practice for many years to use safety relay modules in most basic machine safety applications and it has led to the growth of a wide range of safety-monitoring devices. Essentially these are packaged logic modules certified to be suitable for the standard protection functions. A generic model for these devices is shown in Figure 5.15.

Monitoring relays, E-stop relays and many adaptations of this principle are the workhorses of machinery safety practice and can be found in most machines typically used in manufacturing and automation. Control system equipment vendors offer a wide range of devices selectable for function and for safety category. Examples are shown in Figure 5.16.

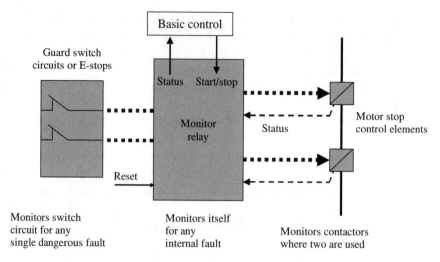

Figure 5.15
Functional diagram of a monitoring relay application

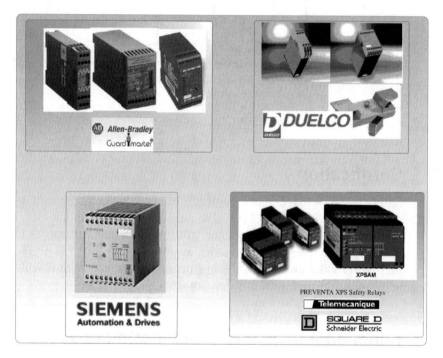

Figure 5.16
Examples of monitoring relay products

The safety relay module provides the following functions:

- Buffers the E-stop switch or any other safety sensor contact from the load-switching duties.
- Provides electrical isolation between the safety switches and the power control circuit.
- Provides multiple output channels for tripping several loads from one set of inputs.
- Performs its safety function even if an earth fault occurs on the input device.

- Provides self-testing each time the unit is operated.
- Uses positively guided contacts in the relays to make sure that contact sets have positive open or closed states.
- Ensures the stop function stays tripped in stop mode after a power break or after any safety trip event.
- Provides for a manual reset action where needed.
- Provides a modular packaged solution for a very common application.

5.7.1   Using monitoring relays with PLCs

With PLCs being the normal basic control device, the safety module interfaces between the output of the PLC and the final control elements (contactors) typically as shown in Figure 5.17.

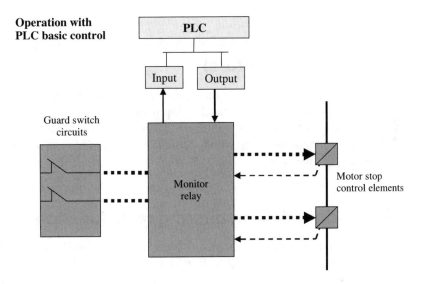

Figure 5.17
Example of functional separation between PLC control and a safety function

This configuration allows normal PLC control functions to be interlocked with a secure hard-wired device performing the safety function. The safety device also provides contact inputs to the PLC to copy the states of sensors and to notify the status of the safety interlocks. The same arrangement is applied for E-stops, presence of detectors using light curtains, pressure-sensitive mats, two-hand controls, guard-locking devices, rundown speed interlocks, etc.

Figure 5.17 also illustrates the point that it is essential to maintain the functional separation between basic controls and safety controls. It is essential to avoid the possibility that a fault that causes an accident is also able to defeat the safety function. Hence it is not acceptable to take away the safety-related control circuits and place the logic in the basic control PLC.

5.8   Electronic and programmable E-stop monitors

At least one manufacturer (PILZ) offers E-stop monitors with electronic outputs or with combination sets of electronic and mechanical outputs. Solid-state outputs are best suited to low-power frequent switching situations. They would presumably be limited to E-stop category 1 applications as discussed at the start of this chapter.

These units qualify for safety categories provided the correct level of proving is built into the electronic output stages. The most effective way to prove this is by frequent self-testing of the output switch by using a short test pulse to internally detect that the output turns off. The pulse is too short to affect the final element and is therefore invisible to the control system.

An all solid-state module would be expected to have a very high reliability but requires an alternative arrangement to provide its internal proving features since it will no longer make use of safety relays with force-guided relay contacts. In line with the established practices for programmable devices in safety systems this device uses a diverse redundant pair of microprocessors for the logic stage and indicates a 'secondary means of isolation' in the case of a detected malfunction in the output stages or in the internal electronics. We shall see more of these methods in Chapter 8. This product is fully certified to IEC 61508 as a programable electronic systems for use in safety applications.

5.9   Using monitoring safety relays for guards (safety gate monitors)

We have seen how the basic requirements for a secure and self-testing arrangement for E-stop functions has led to the development of specialized modules. The same modules are normally offered for use in movable guard applications also known as safety gates. The safety functions for movable gates are very similar to E-stops. If the gate is opened the machine must stop.

5.10   Review of other monitoring relay functions

Within a safety control scheme, E-stop circuits as well as protective-door closures or safety mats and two-hand controls have to be supervised by safety relays.

A number of products can be configured to safeguard a diverse range of hazards. Some of the functions, which can be performed through these ranges of safety modules are:

- Interlocks – to lock gates and to spool valves and isolators
- Interlocking guards where a solenoid lock is operated when the drive contactor has opened or if zero motion has been achieved
- Rotation sensor units and timer units to prevent access until rundown has occurred
- Two-hand control logic functions.

Given below are description of a few devices being put in market by vendors other than famous ones like – PILZ, Guardmaster (Rock well Automation) and Banner. This is to bring home the point that there exists an extensive and competitive market to meet the requirement at 'affordable price' and hence the endeavor should be to fully meet the safety-related requirements.

Moeller Electric range of protective components includes safety relays, command and signaling devices, foot and palm switches, rotary switches and disconnectors, safety position switches and signal towers covers the following:

- *The AT0-ZB and AT4-ZB position safety relays.*
- *AT0-ZBZ safety position switches featuring a holding power of up to 2000 N have been designed specifically for these tasks.*

- *Users have the choice of a magnet-powered version or a spring-powered model additionally providing reliable interlock even if the power fails. In order to open the protective guard in an emergency an auxiliary release mechanism is included.*
- *Where swift reaction with minimal mounting space is needed, the ATR/TS hinge operated and the ATR/ TK hasp operated safety switches are ideal.*
- *Monitoring hinged doors and trap doors they instantly disconnect the power when the door is just slightly opened, even at an angle of 5°.*
- *ESR electronic safety relays from Moeller monitor the wiring in the control system and the de-energiation of the contactors in the power section.*
- *Used as safety relays for emergency-stop applications, the units switch several release current paths either for instantaneous (EN 60204-1 stop category 0) or for time-delayed (stop category 1) interruption of the power supply, once the emergency-stop button is actuated.*
- *If control cables are damaged in a way, which may cause short circuits to exposed conductive parts or to earth, the electronic safety relays react accordingly. They have ability to distinguish different sorts of failures, and the relays switch off immediately, or they prevent reconnection until the fault is rectified.*
- *Dual-channel versions allow short-circuit and cross monitoring to be implemented.*
- *Safety relays suitable for category 4 use feature an additional restart buttonmonitoring function to prevent tampering. This function ensures that the enable signal is only given after the restart button has been released.*
- *If the ESR is used to monitor safety guards, an automatic start function replacing the reset monitoring feature gives users the opportunity to reduce cycle times within the production area while keeping a high safety level.*
- *Often two position switches indicate the status of one door. In these instances Moeller offers an ESR unit with simultaneity check. This relay is able to recognise a fault, if both position switches don't close within 0.5 s of one another. As a re-sult no enable signal for continued production is given.*

Similarly AS-i products from sensing and control expert, IFM Electronic, also offer safety-related control systems as given hereunder. These products are radically different from the safety relay modules we have been looking at because they are bus-connected devices. We shall see these again in Chapter 8 but here it is worth noting that the concept of a pre-approved module designed for specific safety functions has been extended to a networking arrangement. The method of assuring that the devices meet safety category requirements depends on extensive automatic and continuous self-checking of the device as well as its network linkage to the bus master. Any defect detected by the self-checking routines should lead to safe shutdown of the machine function. Here we quote from the manufacturer's publication:

- *IFM's new AS-i Safe modules, which are coloured bright yellow for ease of identification, are fully approved for use in control systems up to Category 4, as defined in EN954-1.*
- *The manufacturer claims that in the AS-i Safe system the safety modules can be freely mixed with standard AS-i modules on the same AS-i cable, thereby avoiding the need for costly additional field wiring to deal with safety-related functions. This will need to be tested for each application. In addition, no special AS-i master is needed, and AS-i Safe products are fully*

compatible with both traditional AS-i installations, and with installations based on the new enhanced AS-i Version 2.1 specification.

- *At the heart of IFM Electronics' AS-i Safe system is the safety monitor module, which is connected to the AS-i cable in the same way as any other module. This is programmed by the user to monitor safety devices, such as emergency-stop pushbuttons and limit switches, which are connected to the same cable. The safety monitor has a worst-case response time of just 35 ms, and offers two safety outputs which are used by the control system in the same way as the outputs from a conventional safety relay. As well as monitoring safety devices on the AS-i bus, the safety monitor module also continually checks for errors and faults. Should a malfunction be detected, it immediately switches the safety outputs to a safe state.*
- *To ensure maximum flexibility, several safety monitors can be connected to the same AS-i cable, each monitoring different combinations of safety devices.*

  In addition to the safety monitor itself, other AS-i Safe modules available from IFM Electronic include emergency-stop pushbuttons with integral AS-i Safe interfaces.

5.10.1 Product guides

Most of the manufacturers and suppliers of safety components offer extensive application guidance in choosing the right kind of product so that the safety devices conform to the relevant codes and standards. This is offered in form of CDs, interactive websites, charts, application notes, etc. The user should not hesitate to use such help because this will guide him to a more informed and appropriate decision. Refer to Appendix A in this manual.

In particular it is important to check that the safety category of each device is indicated by the supplier. Normally the devices will be certified for CE compliance by an approved testing house and certification date will be available. As we have noted earlier types approval is mandatory for CE marking of safety components. If you are building a safety system the approved connection arrangements will simplify the technical file requirements since you will be using a type-approved design. The final responsibility remains however with the design team to see that monitoring relays are correctly applied.

5.11 Conclusions

It is clear that a wide range of modules is available to suit commonly required applications. These modules enable designers to standardize on the safety control arrangements for many types of functional safety applications. The key features are:

- Input circuit combinations can be arranged to suit most E-stop applications.
- Similar input arrangements and internal designs can be provided for other safety devices such as:
    - Movable guards/safety gate monitors
    - Two-hand control relays
    - Area protection devices such as safety mats and light curtains
    - Speed monitors and standstill monitors
    - Timers.

In the workshop we can run through several illustrations of monitoring relays using displays from vendor catalogs. Vendors provide detailed specifications for each available and these are generally available as PDF file downloaded from the vendor websites. Please refer to the website list in this manual.

References

[1] PILZ safety manual pages 127 – 140 (Section 5.3 Emergency Stop Devices).
[2] PILZ product overview: safety relays (download from PILZ).
[3] PILZ PNOZelog safety relays brochure (download from PILZ).
[4] PILZ catalogue files: PNOZ2 GB, PNOZ X1 GB, PNOZX3, PNOZX4.
[5] Banner: Specifier's guide to machine safety (PDF file).
[6] Banner: ES-FL-2A E-stop monitoring relay catalogue (PDF file).
[7] Guardmaster application examples (PDF file).

6

Sensors and devices for machinery protection

6.1 Contents summary

This module looks at the operating principles of a wide range of sensors and protection devices used in machinery protection. To understand the relevance of many sensing devices used in machinery it is essential to first understand the principles of guarding. Many sensors are designed specifically to work in association with guard devices or with safety systems. Others provide a complete guarding function on their own. The content of this module therefore includes:

- Machinery safeguarding methods
- Guarding monitoring and control devices (limit switches, proximity detectors, locking devices, monitoring relays)
- Safety control devices including

    - Hold to run controls
    - Two-hand controls
    - Presence sensing devices
    - Safety mats and edge detectors
    - Opto-electronic devices
    - E-stops.

6.2 Purpose and objectives

The purpose of this module is to ensure that participants are able to recognize and apply the principal types of sensors used in machinery-protection devices. We want to establish the essential characteristics of the sensors and get to understand how they can provide machinery protection on their own or when used with protection systems such as guards or presence sensing controls.

The objective will be to have the ability to select the most appropriate sensor for any given problem bearing in mind the constraints of:

- Meeting the safety requirements
- Finding an affordable solution
- Achieving the highest possible efficiency and productivity for the machine
- Maintaining the safety performance.

6.2.1 Introduction

Most sensors used in machinery protection have been developed specifically to work with mechanical guarding systems or to provide direct electronic guarding systems such as light curtains. We therefore need to begin by identifying the basic types of guarding and protection schemes used in machinery safety. From the essential features of each type of protection we shall see the type of sensors that are needed to support the functions.

We are principally concerned here with the protection of persons from harm due to the movements or actions of the machines. There is also the need to safeguard persons from harm due to electrical hazards. This subject is not covered in this workshop but the widely accepted standard IEC 60204-1 defines the requirements for this aspect of protection.

The basic ways of avoiding injury to persons who could be exposed to danger areas of machinery are:

* Keep persons away from the danger areas whilst the machine can move or
* Prevent the machine from moving when a person is in the danger area or
* Do both of the above.

Ideally the dangerous parts of a machine should be securely fenced to keep persons away from harm. This includes the potentially dangerous tasks of setting up and maintenance testing. The problem, of course, is to leave enough convenient and safe access to make sure the production job gets done!

Safeguarding types, therefore, are selected to balance the needs of cost, practicality, productivity and safety. The performance of each type of safeguard is strongly supported by the sensing devices.

6.2.2 Safeguarding types

The major subdivision of safeguards is between 'guards' and 'safety devices'. Both of these are to protect persons from the hazards that cannot reasonably be removed or be sufficiently limited by design.

Within each type of safeguarding there are recognized types of protection. In standards EN 292-1 and EN 292-2 each type of safeguard is defined.

Guards are defined in EN 292-1 as,

Part of a machine specifically used to provide protection by means of a physical barrier. Depending on its construction, a guard may be called casing, cover, screen, door, enclosing guard, etc.

Safety devices are defined in EN 292-1 as,

Device (other than guard) which eliminates or reduces risk, alone or associated with a guard.

A safety device should be provided such that it automatically prevents the operator from coming into contact with the dangerous part of the machinery. A safety device has to ensure that:

* Whilst the dangerous part is exposed it does not move
* The motion of a dangerous part will be stopped before any person can reach it.

The easiest way to see the different types is to have a structure diagram such as Figure 6.1.

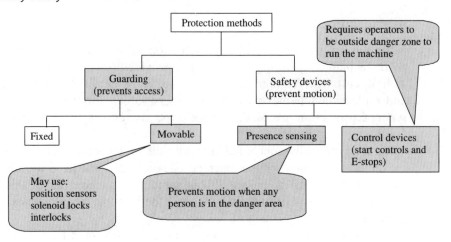

Figure 6.1
Division of protection methods

6.2.3   Guards

All guards serve to prevent physical access to the danger areas. Within the grouping of guards there are a range of devices from fixed guards with no electrical devices through to guards that are movable, power-assisted and have high-integrity limit switches interlocked to the drives of machines.

6.2.4   Safety devices

Safety devices work in a variety of ways to prevent machinery movement if there are persons in the danger area. Therefore these devices employ various types of sensing devices to either detect persons in the danger zone or to confirm they are definitely not in the danger zone.

6.2.5   Distinction between control devices and presence sensing trip devices

Our diagram shows an important distinction between presence sensing trip devices and control devices. Control devices are start/run controls that are positioned outside of the danger area, thus compelling the operator to be clear of the danger area when he or she starts the controls and keeps the machine running.

   Presence sensing devices will detect any and every person in the danger area and are normally interlocked to the machine controls to stop all motion when detection occurs. Logically presence sensing can provide better risk reduction and greater flexibility for production than control devices but this depends on the reliability of the presence sensing system and its ability to detect persons or parts of persons. Control devices are generally less expensive to install than presence sensors.

6.2.6   Relationship between sensors and protection types

It is useful to show the types of sensors associated with the protection types shown in Figure 6.1. We can expand the diagram as shown in Figure 6.2 for guards and Figure 6.3 for safety devices.

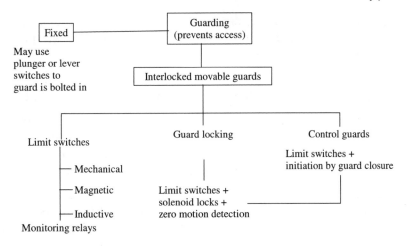

Figure 6.2
Protection methods and sensor types for guards

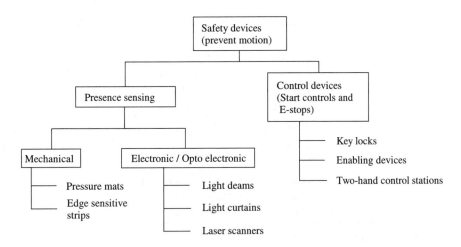

Figure 6.3
Protection methods and sensor types for safety devices

The above diagrams help to identify the roles of the commonly encountered sensors used in machinery safety. The technical features of the sensors are strongly influenced by their roles in providing safety.

6.2.7   Selection of protection method

The procedure for selection of a particular protection method will be examined further in Module 7. A basic guide to selection is set down in standard EN 292-2 and can be used for reference at this stage if needed.

6.2.8   Identification of sensor types

With the aid of Figures 6.2 and 6.3 it becomes possible to define the following list of sensors and actuator types commonly used in machinery protection (see Table 6.1). Please note that this list is based on devices used regularly for protection purposes and that many other types of sensors will be found in the basic control system applications in machinery. For example, precision machine tools may use position measurement systems such as linear and shaft encoders, injection molding machines require pressure and temperature transducers.

Device Type	Operating Principles	Safety Application
Limit switch	Mechanical: lever or plunger operated Mechanical: tongue operated Magnetic: coded magnet Inductive proximity	Guard position detection, for power or control interlocking guards Safe position detection for movable parts
Safety key switch	Mechanical key switch with limit switch mechanism	Guard control device Machine start interlock
Trapped key switches	Mechanical only or mechanical with electrical contacts to main power	Sequence of actions can be enforced by key exchange boxes. Keys provide access to hazardous parts in controlled order
Solenoid lock	Solenoid releases handle to guard or releases trapped key for use on guard. Solenoids controlled by power and motion sensing devices	Prevents opening of guard Prevents removal of trapped keys until safe
Linear or rotary motion detector	Inductive pick-up detecting moving parts Alternative: control unit sensing regenerated voltage from drive motor until stationary	Output applied to guard locked shut by solenoid until motion stops
Enabling control switch Hold to run control device	Mechanical with electrical contacts. Located safely away from the hazard	Control device Switch that must be held in position to allow a machine to start Additional to start control
Two-hand control device	Mechanical or opto-electronic with electrical contacts. Located safely away from the hazard	Control device As above but control circuits require two separate units both actuated to allow machine to run
Trip switch	Limit switch with extended actuator or wire	Contact sensing Trips out drives when physical contact occurs. Mounted close to hazardous parts
Pressure-sensitive mats	Embedded contact plates in floor mats	Presence sensing. Area guarding Trips machine when person stands anywhere on the mat
Pressure-sensitive strips	Embedded contact plates in flexible strip	Presence sensing Stops moving parts when edge contacts an object or person. Also used for area guarding on fences

Opto-electronic device Alt name. AOPD: Active opto-electronic protective device (EN 61496)	Infrared pulse beam source and detector set with control unit. Single or multiple beam sets Multiple beam arrays are called light curtains when they protect against hand or arm intrusion	Presence sensing Point of operation guarding (presence of finger/hand at point of operation) Area of operation guarding (presence of arm/leg in the area of a hazard) Access control guarding (entry of a person to a hazardous area through an access point)
Opto-electronic device Alt. Name AOPDDR: active opto-electronic protective device responsive to diffuse reflection (EN 61496)	Laser light pulses reflected from object in narrow beam. Distance computed from time of flight Beam direction scans around area of protection	Presence sensing Access control guarding (entry of a person to a hazardous area through an access point) Auto-guided vehicles/forklift trucks

Table 6.1
Summary of sensor types and safety devices commonly used in machinery protection

Now that we have a view of most sensors and protection systems we will look more closely at the requirements of each type of protection and see how the sensors have been designed to meet these needs. The next section looks at guarding principles and the sensors they use.

To deal with this subject efficiently we can divide the sensors into some basic types and then consider three questions for each type of device:

How does it work?	The operating principle and its limitations
How does it become fail-safe?	How does the device or subsystem achieve the fail-safe for self-diagnostic characteristics that are essential for safety applications?
Where is it used?	The type of applications it serves

The range of devices we are going to examine consists of:

- E-stops
- Limit switches and proximity switches for guarding
- Guard interlocking devices
- Trapped key systems for access control
- Presence sensing devices such as two-hand controls, safety mats and other tripping devices
- Light curtains for point of operation guarding, area guarding and access guarding
- Laser scanners for area protection and mobile equipment.

This chapter includes brief outlines of the common applications of sensors including opto-electronic devices. For a more detailed study of light curtain application methods such as muting and 'single-break/double-break' operating modes please see Chapter 7.

6.3 Review of guards

Here we review guards as protection systems and then examine the sensors we have already identified as being integral to guard applications. We shall briefly introduce:

- Fixed guards
- Movable guards
- Adjustable guards
- Interlocking guards and locks
- Control guards.

6.3.1 Design requirements for guards

Guards and barriers should be designed so that people cannot reach over, around or through them and come into contact with the prime movers, transmissions and other dangerous parts of machinery. Some other criteria, which should be followed, are as hereunder:

- It should offer positive protection to the operator under all normal working conditions.
- It should be robust, sturdy and rigid and should be recognized as an integral part of the machine.
- Not give rise to any additional risk or danger by itself.
- It should offer protection to others nearby.
- Not be easy to bypass or render non-operational (fixed enclosing guard).
- Be located an adequate distance away from the danger zone (fixed distance guard).
- Cause minimum obstruction, enabling essential work like – oiling, adjusting, etc., to be carried out without dismantling the guard.

The ergonomics in designing of machine guarding must be considered while designing such guards and barriers. These design criteria are generally based on the physical dimensions and reach capabilities of humans. It is essential to take the reach capabilities of the people into account when designing guards and barriers. Hence we find a series of type B standards exist defining the minimum safety distances for fingers, hands, arms and legs (see the next section).

6.3.2 Fixed guards

EN 292-1 defines Fixed Guards as,

These guards are fixed in place; either permanently i.e. welded or fastened, and can only be removed with the aid of tools (i.e. not with a coin or nail file).

Features: These guards are barriers fixed on the machine. They should be constructed and arranged so that it is impossible for any person to reach the dangerous parts. This includes access from the rear or the sides of the machine. They should be fastened securely. It is not suitable to secure fixed guards with butterfly nuts or similar arrangements that allow the guard to be removed by hand.

Typical applications: Generally used to cover belt drives, conveyors, pulleys and drives, chains, gear box flywheels, rotating shafts and rollers, etc. (see Figure 6.4).

Advantage: It is less costly, needs less maintenance and periodic attention.

Disadvantage: Not suitable where frequent access to guarded area is needed.

Where possible, fixed guards should not be able to remain in place if the fixings are removed (i.e. it should not be possible to lean the guard in order to cover the danger zone). A fixed guard may be the simplest of all the protection devices, but there are still some important aspects to consider in their application.

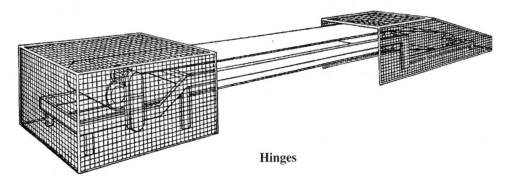

Hinges

Figure 6.4
Typical guard for a head and tail pulley

Let us look at some examples where these are generally used in industry for machinery guarding (see Figure 6.5).

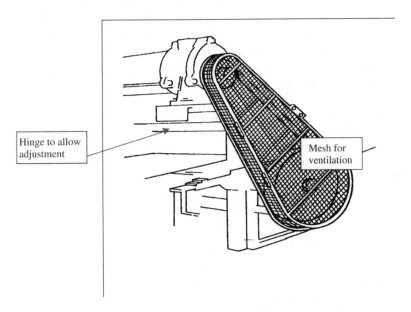

Hinge to allow adjustment

Mesh for ventilation

Figure 6.5
Enclosing guard constructed of wire mesh and angle section preventing access to transmission machinery

In Europe the following EN standards are to be taken into consideration when designing these guards:

- EN 953 (Safety of machinery. Guards. General requirements for the design and construction of fixed and movable guards). This is the starting point. This specification will describe such things as guard height, mechanical requirements and fixings.
- EN 294 (Safety of machinery. Safety distances to prevent danger zones being reached by the upper limbs).

- EN 349 (Safety of machinery. Minimum gaps to avoid crushing of parts of the human body).
- EN 811 (Safety of machinery. Safety distances to prevent danger zones being reached by the lower limbs).

6.3.3   Distance guards

These prevent access to dangerous area through a barrier or fence. These ensure safe distance between dangerous or hazardous area of machinery and persons either operating it or present in the plant.

6.3.4   Adjustable guards

Adjustable guards are used to allow access only to those areas where it is strictly necessary. It should be possible to adjust these guards both manually and automatically, without the use of tools. Where adjustable guards are required, operators should have access to other protective devices such as jigs or push sticks, for example.

6.3.5   Safe by position

This method of guarding relies on dangerous machinery parts being out of the reach of people. The deficiency with this method of making machinery safe is that often the dangerous parts do become accessible when people use ladders and other things to gain access to the dangerous parts. Safe by position is only a suitable option when policies and practices are set in place by employers to ensure that the protection provided by this method is not compromised.

6.3.6   Automatic push-away guards

With this type of guarding a barrier moves toward the machine operator when the hazardous part of the machine operation occurs. This in turn requires the operator to step back out of reach of the hazard. It should be noted that if this type of guard is not carefully designed and maintained then the push-away guard can itself become a hazard and cause injury. Thorough training of operators in the safe use of machinery guarded in this way is essential.

Let us look an example how exposed rotating cutting machinery including *cut-off saws, milling machines, and friction cutting and boring* equipment is protected. Figure 6.6 gives an idea how fixed and moving guards should be fitted where appropriate.

A particular point to note is the self-actuating visor fitted to the fixed guard. If this visor is not affixed, the cutter's teeth are exposed when the machine is at the top of its stroke. As the cutter is lowered the visor automatically rises.

On a larger scale, push-away guards are often found on large press tools and shears where a horizontal bar sweeps away the person who is standing too close to the tool or blade as it descends.

6.3.7   Movable guards and interlocking guards

Movable guards are the sorts we see frequently on machines having hinged covers or sliding panels over working parts. These are defined in EN 292-1:

Guard generally connected by mechanical means (e.g. hinges or slides) to the machine frame or an adjacent element and which can be opened without the use of tools.

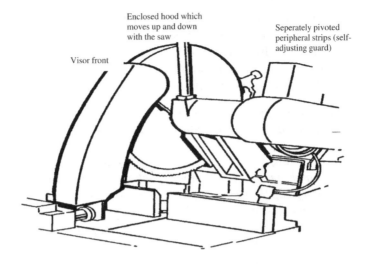

Figure 6.6
Self-adjusting guarding arrangement for a cut-off saw

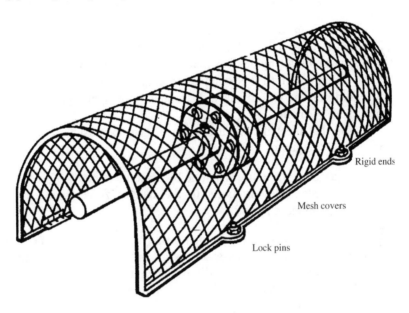

Figure 6.7
Type of a movable guard used to guard all nip-points in case of pulleys and drives, so as to be out of reach for personnel

The above implies that it should remain fixed to the machinery when open and absence or failure of one of its components must prevent start-up (see Figure 6.7).

More specific classification of guard types is given in EN 292, Part 2, Section 4.2. In paragraph 4.2.2.3 the standard describes the requirements for *type a movable guards* as:

Movable guards against hazards generated by moving transmission parts shall:

- *as far as possible remain fixed to the machinery (generally by means of hinges or guides) when open,*
- *be interlocking guards with or without guard locking in order to prevent moving parts starting up as long as these parts can be reached and to give a stop command when they are no longer closed.*

To comply with the requirements of type a, the guard switch or switches will normally operate in the same way as an E-stop function. This type of interlocking is called *power interlocking*. It is likely to be used for situations of infrequent access such as adjustment of transmissions or for lubrication.

EN 292-2 describes the requirements for *type b movable guards* as:

Movable guards against hazards generated by other moving parts shall be designed and associated with the machine control system so that:

- *moving parts cannot start up while they are within the operator's reach and the operator cannot reach moving parts once they have started up; this can be achieved by interlocking guards without guard locking or with guard locking.*
- *they can be adjusted only by means of an intentional action, such as the use of a tool, key, etc.*
- *the absence or failure of one of their components prevents starting or stopping the moving parts; this can be achieved by automatic monitoring.*
- *protection against ejection hazards is ensured by appropriate means.*

It is significant that the standard says: 'this can be achieved by automatic monitoring'. This invites us to use the fail-safe and self-checking features of guard switches combined with guard monitoring devices that we introduced in Module 5.

The function of the associated locking device may be more or less sophisticated, depending on the type of hazard, frequency of opening, etc. This will be determined by the risk assessment.

Guards that meet the requirements of type b must be considered carefully. Does the opening of the guard:

(a) Stop the entire machine by disconnecting the power? (power interlocking).
(b) Stop moving parts in the danger zone, guarded for the duration of this opening period?

To comply with the requirements of (b), the method and integrity of the guarding control circuit has to be assessed as an individual item and the relevant specifications consulted. A risk assessment will also have to be performed. This type of interlocking is called *control interlocking*. EN 292-1 summarizes the functions of control interlocking guards as follows.

These guards are associated with interlocking controls, so that:

- *The hazardous machine functions ''covered'' by the guard cannot operate until the guard is closed,*
- *If the guard is opened while hazardous machine functions are operating, a stop instruction is given,*
- *When the guard is closed, the hazardous machine functions ''covered'' by the guard can operate, but the closure of the guard does not by itself initiate their operation.*

6.3.8   Features of interlocking guards

In principle, guard interlocking links functions of operator and machine in such a way that parts are not accessible when danger is present and where parts are accessible, the danger is eliminated. Interlocked guards may operate mechanically, hydraulically, centrifugally, pneumatically, electrically (or a combination of these ways), and make the machinery safe by ensuring that the hazard is not present when the guard is opened. Where there is likely to be overrun on a machine then devices which do not allow the guard to be opened until the motion of the machine has stopped or devices which will stop the motion of the machine, should be fitted. It is essential that any brakes fitted to machinery be well maintained.

Interlocked, rise and fall guards or machine tools which are capable of inflicting injury if they fall freely under gravity should be provided with a suitable anti-free-fall device. Power-operated guards should be operated with the minimum of force to prevent the guard from causing a trapping hazard. Where it is not possible to reduce the closing force of a guard, a safety trip device could be fitted to the leading edge of the guard, which will stop and reverse the motion of the guard.

If the guard is rarely opened, it is recommended that the functioning of the interlock switch should be checked (with the guard closed) on a regular basis (start of shift or daily) to enable the safety relays to detect single faults.

Typical applications: Generally used for sliding or hinged doors interlocked with drive, cam-activated switches, trapped key systems, mechanical interlocks, etc.

Advantages: Accessibility to guarded areas when there are no hazards present is ensured and hence more productive. Can be built into machine design and can include ejection and noise protection. Example: CNC lathes and machining centers.

Disadvantages:

- In some cases: time-consuming to operate.
- Expensive and cumbersome for large areas and large machines.
- Where this problem become severe it may be attractive to consider opto-electronic guarding.

Electrical control interlocking

Electrical control interlocks linked to guard positions are common where rapid or frequent access is required into a machine. Power to the machine control system can be maintained while providing a safe method of entry. With the guards open, the interlocks prevent movement of hazardous parts by blocking the control circuits of the final drive contactors. Guard positions are detected by safety-rated limit switches usually interfaced to the controls via guard switch-monitoring relay systems that can diagnose circuit defects and provide safe failure modes.

Gate control is provided by means of solenoid-controlled locks that also contain safety-monitoring circuits.

These circuits incorporate positively guided contacts that control the solenoid locks and monitor the physical position of the gate. Additional electrical contacts are provided to help determine the machine status. Tongue entry or magnetic guard switches are typically used on sliding doors, while handle-operated products can be used for hinged gates.

Summarizing here: We have seen related types of guarding systems:

- Power interlocks based on guard position
- Control interlocks based on guard position
- Locking guards with control interlocking.

Now let us consider the sensing and control requirements.

6.4 Sensing devices for guards

Interlocking guards demand a special type of position limit switch sensor to ensure that they are always closed before the safety circuit will release the interlock on the power to the machine.

- Limit switches must be sufficiently accurate to detect guard position within defined ranges.
- Tolerant enough to accept some misalignment as wear develops.
- Self-aligning to a limited extent.
- Tamper-proof to prevent operators trying to jam the switch closed and operate without the guard in position.
- Reliable in respect of the contacts always following the movement of the actuator. We need positively guided contacts or designs that avoid internal linkage.
- Fail-safe operation and ability to meet safety integrity requirements (safety categories).
- A large range of sizes and power ratings to suit applications.
- High degree of immunity to dust and liquids (IP67 ratings).

There are a large number for product types available to the designer. Here we try to identify the main types and their features.

6.4.1 Mechanical limit switches

With these devices the guard door is linked mechanically to the contacts of the switch using positive mode operation in compliance with standards and as applicable to E-stop switches. There are three main types of mechanical actuation. These are:

1. Tongue-operated
2. Hinge-operated
3. Cam/plunger-operated.

Tongue-operated limit switches

Features: These devices comprise two separate elements: a switch body and actuator tongue. Tongues are metal probes specially shaped to fit into the switch rather like a key. These are to be fitted to edge of a sliding door or on to a removable guard. When the tongue enters the switch body it engages a mechanism that closes or opens internal electrical contacts. The mechanism is designed to prevent easy bypassing or 'cheating' of the switch since the tongue is coded like a key. Usually these types have self-ejecting spring-loaded mechanisms so that the tongue will only remain in place if it is attached to the door of the guard. If the tongue were to be removed from the door and just pushed into the switch it would not stay in place.

- Requires reasonably accurate alignment
- Should not be subjected to constant high-amplitude vibration
- Can be used on sliding, hinged and lift-off guards.

Advantages: Low cost, versatile, certified for safety. Almost tamper-proof if the tongue design is good.

Disadvantages: Not suited to pharmaceutical applications and some food applications where good cleaning is essential.

Hinge-actuated limit switches

The device is mounted over the hinge-pin of a hinged guard. The opening of the guard is transmitted via a positive mode operating mechanism to the control circuit contacts.

Advantages: When properly installed these types of switches are ideal for most hinged guard doors where there is access to the hinge centerline. They can isolate the control circuit within 3° of guard movement and they are extremely difficult to defeat without dismantling the guard.

Disadvantages: Care must be taken on large wide guard doors as an opening movement of only 3° can still result in a significant gap at the opening edge on very wide guard doors. It is also important to ensure that a heavy guard does not put undue strain on the switch actuator shaft.

Cam-operated limit switches

This type of arrangement usually takes the form of a positive mode acting limit (or position) switch and a linear or rotary cam. It is usually used on sliding guards and when the guard is opened the cam forces the plunger down to open the control circuit contacts.

Features: The simplicity of the system enables the switch to be both small and reliable. It is extremely important that the switch plunger can only extend when the guard is fully closed. This means that it may be necessary to fit stops to limit the guard movement in both directions.

Advantages: Wide range of low-cost switches are available. Available in very wide range of sizes. Can be made extremely durable and rugged. Easy for maintenance crews to inspect and repair.

Disadvantages: Relatively easy to defeat. It cannot be used on lift-off guards. Requires careful installation and design of strikers: For example: it is necessary to fabricate a suitably profiled cam which must operate within defined tolerances. This system can be prone to failures due to wear especially in the presence of abrasive materials or with badly profiled cams.

6.4.2 Non-contact (non-mechanical) actuation

With these devices the guard door is linked to the control circuit contacts of the switch via a magnetic or electronic field. To make these types of devices suitable for interlocking duties they must incorporate enhancements to ensure their satisfactory operation.

Magnetically actuated switches have the advantage of being completely sealed and the profiles of actuator and switch body are plain. These will be preferred in food and pharmaceutical applications where hygienic and easy-to-clean devices are essential. The measures taken to avoid tampering are based on multiple poled magnet strips that are uniquely coded for the matching switch. The switch sensor is a sensitive reed switch that must be activated in the right polarity to achieve a valid circuit. There is no possibility that a simple magnet can be used to override the switch.

Reed switches have low contact ratings, so the devices are protected by fuse links but should be installed with monitoring relays to ensure overloads do not occur.

Here are some examples quoted from manufacturer's publicity notes:

Guardmaster has introduced new Sipha 6 control unit. It is claimed that the Sipha 6 improves upon the unique functionality of the magnetically coded Sipha system to deliver unmatched levels of non-contact reliability, protection against defeat by tampering and fail safe operation.

Telemecanique has also introduced small, easy to install, and highly tolerant to misalignment, the new XCS-DM range of contact less coded magnetic safety switches. The manufacturer claims that they are ideal for ensuring that dangerous moving machinery is stopped as soon as the guard is opened. All XCS-DM switches use coded reed technology, which provides effective protection against tampering, as actuation with a simple magnet is impossible.

We should note from these examples of the manufacturer's claims that the only product being offered here is the sensor. The ability to stop the machine, of course, depends on the design of the complete safety-related system. The manufacturer is really only claiming that this device is ideal for use in such an application.

Catalogs

The best way to become familiar with the wide range of limit switches for safety systems is peruse the catalogs of the large manufacturers. The wide range of products is usually displayed with a useful selection guide. In this workshop we shall illustrate many of the switch types by running selection guide software from manufacturers such as PILZ and AB/Guardmaster.

6.4.3 Suggested references

- PILZ interactive safety guide. CD or download from www.pilz.com.
- Rockwell Automation, AB Guardmaster website for products: www.ab.com.

6.4.4 Summary of guard sensor safety issues

Let us see if we can answer our basic safety questions applied to guard sensors.

1. How does it work? See above. Versions can be found for all common machinery situations.
2. How does it achieve the fail-safe and diagnostic performance needed for safety?

All limit switch applications associated with guards have a number of features in common to meet safety requirements:

By design:

- Switch designs are selected to match the mechanical needs of the guard type so that any misalignment or incomplete closing will not actuate the switch.
- Switches must have tamper-proof designs since they are frequently accessible to operators. Tongues are coded and sockets are spring-loaded, hinge types are less accessible, magnetic types have coded multi-pole magnets.
- Switches are designed for positive mode actuation in the same manner as an E-stop switch. The 'shelf' position for the switch will present open contacts, the guard must be in position to operate the actuator and close the contacts.

By redundancy All safety designs for categories 3 and 4 need to be fault-tolerant. In the case of limit switches the faults could be in the interaction with the guard parts or in the switch itself or in the electrical circuits. The only effective way to overcome the mechanical interface faults is to have redundant switches and actuators (see Figures 6.8 and 6.9). For category 4 systems the redundant switches will need diverse arrangements to reduce the chances of a common mode failure.

Single tongue switch on sliding guard with montoring relay.

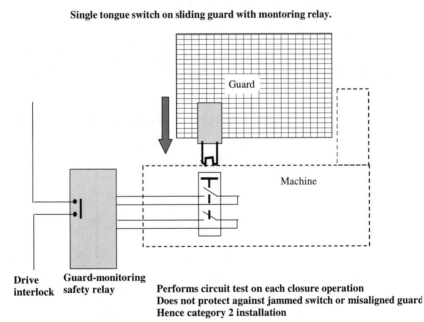

Guard

Machine

Drive interlock **Guard-monitoring safety relay**

Performs circuit test on each closure operation
Does not protect against jammed switch or misaligned guard
Hence category 2 installation

Figure 6.8
Guard with single limit switch

Within each limit switch the normal standard is to have two redundant safety contacts. These will match up with categories 2, 3 or 4 rated electrical circuits, which are best provided by using guard-monitoring safety relays. Please refer back to Module 5 for details of safety relay principles.

Many guard switch applications will want to protect against misalignment of the guard or the possible malfunction of the switch. Hence two or more switches are often fitted at different points on the guards. Where risk assessment indicates the need for category 3 or 4 protection the use of redundant limit switches is essential.

Two tongue switches on sliding guard with monitoring relay.

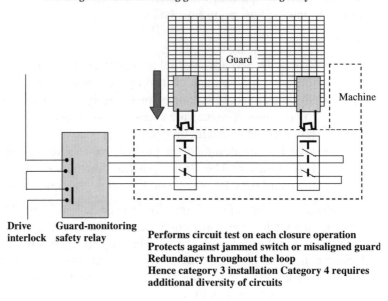

Drive Guard-monitoring
interlock safety relay

Performs circuit test on each closure operation
Protects against jammed switch or misaligned guard
Redundancy throughout the loop
Hence category 3 installation Category 4 requires
additional diversity of circuits

Figure 6.9
Redundant guard switch arrangement for category 3

Another common practice with guard switches is the use of redundant and diverse arrangements to eliminate most chances of defeat by operators or to avoid common mode failures. On a sliding gate, for example, the arrangement depicted in Figure 6.10 may be used.

High-integrity switch arrangement for sliding guard

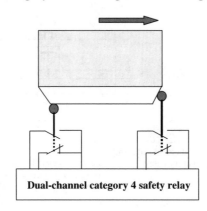

Dual-channel category 4 safety relay

Both switches are monitored for circuit faults. Diverse circuits apply.
Diverse mode of switches apply. Protects against overtravel

Figure 6.10
Sliding gate position detection, category 4 system

6.4.5   Guard-locking systems and devices

The use of locked guards and gates has traditionally been the most reliable method of guarding. The method relies on a responsible person having possession of the key to the guard at all times. This person also being responsible for ensuring that the gate is not opened until the machine is switched off, isolated and all motion of the machine has stopped. Isolation, hold cards and lockout devices are also used to ensure that the machine is not accidentally restarted. The guard or gate is to be locked by the responsible person and the hold cards and lockout devices removed, before the machine is restarted. These are dependent a lot on rigid following of rules and systems and are quite time-consuming.

Locking systems can be divided into two basic types:

1. Mechanical trapped key interlocking
2. Electrical control interlocking.

Trapped key interlocking is a proven high-integrity safety system that complies with the design principles identified in EN 954-1, EN 1088, EN 292-1 and EN 1050. All energy sources (e.g. electrical, pneumatic, hydraulic) can be reduced to zero, providing unrivaled operator protection. Such a system is also very easy to retrofit and can be customized to individual applications. Control interlocking offers rapid access, machine diagnostics, ease of maintenance and the ability to maintain power to the PLC.

When determining the interlocks that should be employed in any key interlock system, it is necessary that someone establish

- The normal state of the controlled devices
- All the possible scenarios
- Just what is to be achieved with the interlock system?

When required, extremely complex interlock systems can be built from basic interlocking principles and applications.

For particularly sensitive operations such as when work needs to be performed under more-than-usual supervision, an interlock system may include a 'supervisory key'. Inserting the normal key alone will not operate the interlock to permit further steps. Instead, a second key (assigned to a designated supervisor) must also be introduced for operation in the same interlock. A 'supervisory key' can be introduced at any point of an interlock system regardless of the complexity.

Although key interlock systems are mechanical in nature, there are electrical accessories available that are quite effective when designing an interlocking system. It is often necessary to make sure a key is held for a certain length of time following shutdown of a controlled device or until a specific circuit is energized allowing the interlocking scheme to be initiated,

- Solenoid Key Release Units and
- Time Delay Key Release Units
  … Are designed for these purposes.

6.4.6   Guard-locking interlock + limit switches

Guard-locking switches can be used for applications where the dangerous motion takes time to rundown after the electrical power has been stopped.

In these applications a guard door can be locked in position automatically on closure of the guard. Limit switches within the lock confirm the lock has been closed. It can be kept locked by the safety control system using a solenoid actuator until it is safe to open the guard again. The guard door cannot be unlocked until the hazardous motion has been stopped. It may also be conditional on time and movement sensing. Unlocking and opening the guard door also de-actuates the position contacts.

The tongue-operated limit switch is able to perform this duty as the tongue is shaped as a latching device. It cannot be removed from the switch until the internal solenoid is energized. In a power fail situation a mechanical override or key release can be used.

One of the most common examples is the lock on a washing machine door. They can also prevent machine damage and programmable control problems caused by unintended stopping in mid-cycle caused when an interlocked guard is opened.

For an illustration of a typical guard-locking switch, see Figure 6.11.

Figure 6.11
Typical guard-locking switch with choice of entry ports and internal solenoid-operated lock

6.4.7   Safety issues for guard locks

The treatment of guard locks is the same as for limit switches. Dual circuits are used for the limit switch that confirms the lock tongue is in place. Safety relays can be used to assist with interfacing to the control system. Monitoring contacts are provided for the solenoid action and for external signaling to show if the lock is on or off. The general arrangement can be seen in Figure 6.12 from Guardmaster but note that this figure does not use a monitoring safety relay to interface to the machine controls. In most safety applications a guard-monitoring relay system would add security to the design and allow it to qualify for safety category 3 if one lock switch is used since the circuit with be single fault-tolerant and tested at each operation. This assumes a low level of faults in the lock and limit switch assembly. A more secure design might consider a dual-redundant guard-locking switch.

Note in the diagram that interlocks contacts K1 and K2 are auxiliaries from the power control contactors. Hence no power can be applied to release the lock until the contactors have de-energized. Guard-locking switches operate on the principle of 'energize to release'.

It is easy to see that if we place another contact in series with K1 and K2 operating from a speed sensor or a time delay the lock cannot be released until safe conditions have been reached even if the machine has a rundown period.

Versions of guard locks are available with a trapped key facility. In operation (machine running), the key is trapped in the lock until a signal is applied to the internal solenoid. When the solenoid is energized the guard door can be opened and the key can be removed, locking the switch in an open state. This of course gives security to the person holding the key who is now also free to enter the area protected by the guard.

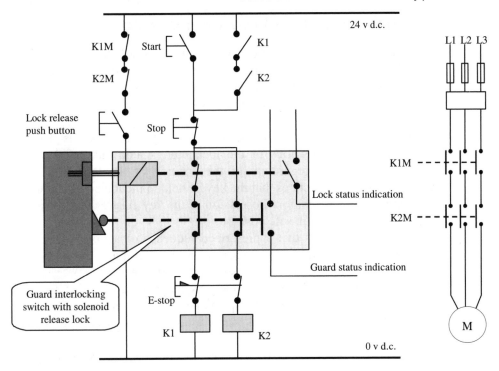

Figure 6.12
Circuit arrangement for a locked guard with interlock from final contactor

6.4.8   Sensors for motion detection

Having described guard interlocks this is good point to briefly describe 'stopped or slow motion detectors'. There are two basic types: inductive pick-up and back e.m.f. Optical types may also be available. Alternatively for drives with consistent run down times adjustable delay timer modules are available.

Inductive types have proximity detectors that generate a pulse each time a projection on a shaft or linear slide passes across its face. The presence of pulses causes the control unit to keep its output contact de-energized. Pulses must be absent for a defined period before the measured item is considered to be stationery at which point the control unit closes its output contacts.

Back e.m.f. versions are connected across motor terminals and detect the back e.m.f. generated by the drive until it comes to rest.

Safety design reminder

All types of motion control units must be designed for fail-safe operation and are triggered to respond when the initial stop command is detected in the power control circuits. These units are an integral part of a safety-related control system and the required integrity rating of the speed interlock function should be determined by risk assessment.

One essential feature for the proximity types is to have a pair of pick-up heads. This provides a good level of assurance that the system can tolerate one head being misaligned. If possible the head should be arranged so that one of them is always reading positive. Hence two negative readings will signal a fault.

6.5   Mechanical trapped key interlocking

A typical trapped keyed interlock is comprised of a lock cylinder, a support housing, a movable locking bolt, and a cam arranged to move the bolt in response to operation of the correct key. Various styles of interlock housings are available and each style is designed to mount in a different way depending upon the equipment to which the interlock is to be installed.

One of the most important features of a keyed interlock is that the key cannot be removed from all positions of the lock bolt. A conventional lockset may allow free removal of the key regardless of the position of the lock bolt. The function of an interlock, however, dictates that the key be held in the lock cylinder unless the lock bolt is in a predetermined position. Possession of the key ensures that the associated device has been locked in a known, safe position.

This system ensures that a prescribed sequence of actions is taken when accessing a machine. It is of particular use where there are multiple hazard types or where access is required to a number of danger zones over a wide area.

The principle behind mechanical key exchange control is that all sources of power are isolated and all stored energy dissipated before the hazardous area of the machine can be accessed. This tried and tested methodology can be used on all machine installation categories.

A number of products can be configured to safeguard a diverse range of hazards:

- Interlocks can be used to lock gates and to lock hydraulic spool valves and electrical isolators.
- They can also be used to ensure that sources of stored energy are made safe, for example, gas cylinders.

Locks are designed in such a way that the key can only be removed when the hazard has been isolated and can only be reinstated when the key is trapped in the lock. This means that the key represents the hazard status associated with that lock. Keys are uniquely coded and can therefore control the sequence and limit access to authorized personnel.

Solenoid-controlled locks ensure that a key is trapped until signaled by another action. This could be a permission signal from a remote source or it could be part of the machine shutdown system.

As already mentioned, time delay units can be used where a machine has a rundown period. Under this system it is not possible to take out the key until power has been removed and the pre-set time period has elapsed. If the rundown period is predictable, the machine can be presumed to have come to rest and the key can be used in the next sequence of machine access.

Rotation sensor units operate in a similar way to time delay units, but use measurements to prove that the rotating part of a machine has stopped before access is granted.

Key exchange boxes can be used to ensure that certain actions are performed before others. They also allow complex 'if/or' sequences to be safely controlled. Figure 6.13 shows an example of a key exchange box.

A safety key is an important feature of mechanical trapped key systems. The key is removed and taken into the hazardous area, ensuring that a machine cannot start-up unexpectedly. This is particularly important where personnel can move out of sight within a guarded area. Maintenance personnel can therefore have uniquely coded or sub-master keys, ensuring that only suitably trained staff can instigate access.

Figure 6.13
KIRK® key interlock systems: key exchange box

The two systems can also be combined so that safety keys can be used to protect individuals, while access keys are used to limit access to authorized personnel. This is particularly useful when a robot needs to be put into 'teach mode' or a machine has to be reset.

In the keyed interlock systems, a group or series of interlocking devices is applied to associated equipment in such a manner as to prevent, or allow, operation of the equipment in a prearranged sequence only.

Example of a key interlock systems application

An innovation developed by Mistura Systems known as '*a search and seal*' interlock system to prevent personnel from becoming trapped in potentially hazardous areas. Operatives such as maintenance engineers, cleaners and technicians often need to gain access to large protected zones like automatic warehouses, cold stores, industrial boiler filters and X-ray cells. Such areas must be completely free of personnel before they are sealed and re-energized. The Mistura Search and Seal system ensures this.

- Once a search of the area has been completed, a safety key must be taken to the furthest point in the area and inserted into a spring-loaded switch.
- This activates a warning siren, flashing lamps and a timer circuit, commencing the lock-up procedure. The safety key must then be taken directly to the exit point and inserted into the door lock.
- The door bolt is then engaged, enabling the control key to be removed.
- The area is now secure.
- The control key must be inserted into the system panel, mounted on an adjacent wall, within a specified time period.
- If it is outside the time period a search of the area must be carried out again and the sequence repeated.

In summary we can see that trapped key interlocking supports any safety system where complex access procedures must be followed at relatively infrequent intervals. It ensures that correct procedures are followed even if the persons involved cannot recall the exact procedures at the time of use.

Limitations of trapped key systems

This type of safety function is unlikely to be suitable for applications where frequent and rapid access to hazardous parts must be made safe for a production operative. Time lost in gaining access, the complexity of actions and the temptation to develop shortcuts all indicate the need for a better method of protection in such cases. This type of situation is often better served by the presence sensing safety devices such as light curtains.

6.6   Presence sensing devices

Mechanical guarding, whether fixed or movable, may not always provide the solution for certain types of machinery applications. When it comes to production lines and large automated manufacturing plants there are major cost and productivity benefits to be achieved if the persons interacting with the machines can be as free as possible to move into the workspaces needed for the production work without hindrance from physical guarding devices. So the alternative approach to guarding is:

- Make sure the presence of a person in the danger zone is always detected and the hazardous motion is stopped (Trip action).
- Make sure the machine cannot start movements until there are no persons in the danger zone (Control action).

Here we look at the sensing devices that support these functions. In the next module we shall look at some of the application techniques for the light curtain versions.

6.6.1   Trip devices

A trip device is a general term for a safety device arranged so that if a person reaches into or approaches a dangerous part of a machine, the trip device will be activated to stop the machine movement.

Recall that there are two basic methods of stopping a machine for safety:

1. Power tripping acts the same way as an E-stop. It interrupts power to all systems in the machine.
2. Control tripping stops all drives and sets all actuators to safe positions but leaves power on the controls of the machine.

The first is simple to use and install but creates problems in complex machines especially where PLCs are involved. There is a risk of lost production or damage to the machine. The second requires safety-related controls to interact with the basic machine controls and hence can be more complex to design. However the benefits are found in the ease with which machines can be stopped and restarted as soon as they are safe.

The trip devices we are going to look at here are:

- Mechanical trip switches
- Safety mats
- Edge detectors
- Opto-electronic devices/light beam and curtains.

Importance of 'safety distance'

When any trip device is activated, it must cause, via a suitable interface, an E-stop of the machine either through a power trip or a control trip. If the trip fails, there is no substantial physical barrier to stop people from coming into contact with the dangerous parts. All safety trips must be evaluated for safety category as we have seen earlier. In many common applications such as in large press brakes the trips must also release braking devices to stop the movements as quickly as possible.

The stopping time of the machine hazard therefore becomes a highly critical factor in the design of the trip system. The longer it takes to stop, the greater the clearance is required between the trip device and the hazard. If we know the stopping time and if we know how fast a person can move an arm or leg or whole body we can determine the distance between the point of detection and the point of contact. This is called the 'safety distance.' We shall look at safety distance calculation in the next module but for an understanding of sensors we need to keep in mind the basic requirements of safety distance.

6.6.2   Mechanical trip switches

Mechanical trip switches are usually seen as limit switches with specially designed actuators. They are connected for power or control tripping in exactly the same way as guard position switches and will often employ monitoring safety relays. These provide a simple protection on any machine where hand or arm contact can be made with the actuator before contact with the hazard.

One of the best examples is the telescopic probe shown here (see Figure 6.14). These devices are usually deployed on machine tools such as drilling machines and the machines are provided with DC injection braking to provide nearly instantaneous stopping. Hence the safety distance can be very small and the probes are adjusted to be close to the workpiece.

Figure 6.14
Telescopic probes

Trip wires

These are usually defined as E-stop systems since they provide a means of activating the E-stop function by grab wires running alongside the danger areas. Again the need for rapid stopping of the machine must be identified. Trip wire switches are specially constructed to ensure they latch out easily when pulled. Monitoring E-stop relays are used to provide the fault detection and feedback monitoring needed to meet category 3 and 4 ratings.

6.6.3 Pressure-sensitive mats

Pressure mats operate on the principle that to gain access to the dangerous part a person has to step on to a mat or platform, which then trips the machine or prevents it starting. Pressure mats may also be used in conjunction with other methods of isolating people from hazards. If used, pressure mats should be of type that will not fail to danger.

Because pressure-sensitive mats do not usually show any external sign of failure, they should be used with a control system that monitors their operation, which in the event of a failure shuts down the machine. Here is an example of features of pressure mat supplied by Grandmaster as per the published literature:

- Hardened steel plate construction
- No dead spots
- 4-wire system to detect opens and shorts
- Can withstand a static force of 4500 psi
- IP67 rated (NEMA 6P).

These mats are controlled and discharge their function in conjunction with the MatGuard Control Units (see Figure 6.15). These units monitor all of the mats, which are connected together to form a safeguarded zone. The safeguarded zone can be up to a total of 100 m^2 and made from any number of mats. The controller is designed to interface with the control circuit of the machine and includes two safety relays to ensure control redundancy.

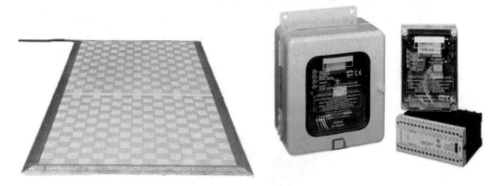

Figure 6.15
Typical safety mat and controller

The controller detects a presence on the mat, a short circuit, or an open circuit. Under each of these conditions, the safety output relays turn off. When interfaced properly, the machine or hazardous motion will receive a stop signal, and an auxiliary output relay turns ON.

Another supplier has a trade name of The CKP/Solo safety mat. This is based on the Tape switch CKP/S1 safety mat, which has been used in industry for many years.

- Safety mat systems normally consist of a safety mat as the sensor and a separate control unit to carry out the safety-monitoring function and to provide high current, normally closed output contacts and a restart interlock. Although this is ideal for most applications it can be unnecessarily expensive when dealing with modern safety bus systems such as the AS-i-Safe interface.
- Safety mat sensors are usually normally open devices. They usually have open contacts when the mat is clear and are closed by the weight of someone standing on the mat.

- This orientation is not suitable for safety circuits and so an additional control unit is necessary to provide the normally closed outputs.
- The external control unit also usually carries out the essential function of checking for short circuits and open circuits in the wiring and sensor.
- The CKP/Solo has built-in electronic circuits to carry out these safety functions and also provides two volt-free outputs that are closed when the mat is clear and open when someone stands on the mat.
- This configuration is ideal for use with ASi-interface systems, which connect sensors and actuators to remote control systems through only a two-wire cable.
- The mats can connect directly to the same kind of standard AS-i-Safe I/O connection modules that are used to connect E-stop switches over the ASi interface.
- This is much more convenient than the alternative of having an additional control unit close to the mat so that it can provide the volt-free, normally closed connections to the AS-i I/O module.

- The design of the switching element of the mat is identical to that of the CKP/S1 mat.
- An array of Tape switch ribbon switches, wired to form a single continuous switch, is sandwiched between two layers of very heavy-duty PVC-based material.

It is claimed by supplier that this is a very durable fail-safe switching mat, which is guaranteed to detect anyone who steps on it.

6.6.4   Edge detectors

Edge detectors find application in many door closure applications, the most obvious example being on elevator doors (see Figure 6.16). In industry they will protect persons from powered sliding doors including those on actuated sliding guards. As with the pressure mat, edge detectors require a fail-safe monitoring control unit (see Figure 6.17).

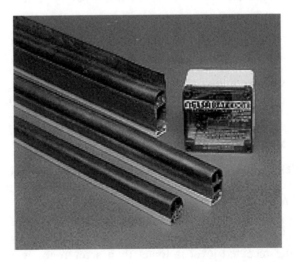

Figure 6.16
Example of edge detector strips, the Nelsa 'safe-edge'

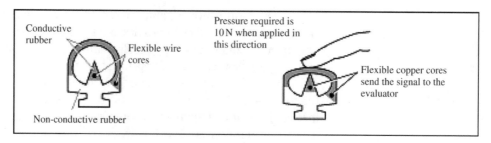

Figure 6.17
Detail of the Nelsa edge detector strip

6.6.5   Opto-electronic presence detectors

Optical guarding and detection systems use light beams in various forms to detect the presence of persons approaching or within a hazardous area. If an operator or maintenance person requires regular access to a hazardous area, there are a number of optical system solutions that are likely to provide the advantages of higher productivity, with protection for both the operator and any third party.

However, it is important to remember that this method of guarding offers no protection against flying materials. An opto-electronic protective device can only be used if the operator is not at risk from injury due to splashes (e.g. molten material) or flying pieces of material.

General principles and types for safety

Optical detection systems have been built in a variety of types but the methods used for safety predominately use the principle of interruption of an infrared light beam between a source and a separately located receiver. These are also known as AOPDs (active opto-electronic protective devices). Reflective beam designs are used in basic industrial controls but for safety systems only the laser beam scanner makes use the reflection of a light beam. These are known as AOPDs responsive to diffuse reflection. (These terms being taken from the widely used standard IEC 61496.)

The recognized types of opto-electronic safety products are as follows:

- Single-beam source and receiver units
- Multiple emitter and receiver sets known as light screens or light curtains
- Laser scanning reflection sensors using time-of-flight measurement of distance.

Here we take a look at the basic characteristics of the opto-electronic sensors and their control units. The application details are rolled over to the next module.

Principles of operation

This description describes a single-channel emitter and receiver pair with an electronic control unit. The principles are the same for each pair in a multi-channel system (see Figure 6.18).

The principle of operation is that an emitter unit sends a narrow beam of infrared light to an optical receiver unit that detects the beam and switches to a 'blocked' or zero state when the received level is lower than a reference level. When the beam is blocked by intrusion in the sensing field, the control unit opens its output contacts to activate a machine trip or E-stop.

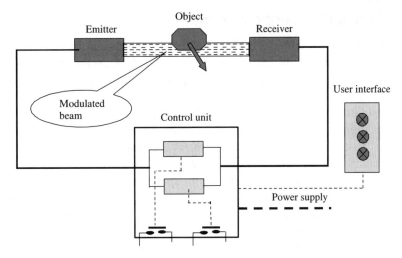

Figure 6.18
Basic components of an active opto-electronic protection device (AOPD)

In order to achieve a high degree of safety reliability, steps have to be taken to avoid false signals and detect faults in the usual manner of categories 2, 3 or 4 safety systems. This is achieved by a number of measures, in particular:

- Very narrow beam with divergence less than typically 1°. This minimizes the possibility of reflection from adjacent objects and helps to avoid cross-talk with any adjacent detectors.
- The light beam from the emitter is modulated so that ambient light levels do not affect the receiver. Applies also to dust on the optics and smoke interference.
- The detection threshold is cyclically tested by pulsing of the emitter to an 'off' state and confirming that the receiving system responds with an 'off' state at the same time.
- Mounting assemblies are designed to absorb vibration from machines to resist disturbances to the beam path. This could be a problem every time an impact or gate closure occurs.
- Fail to safety (off state = blocked path) design of the optical circuits.
- Control units have dual-redundant evaluation and diagnostic sets for categories 3 and 4 safety systems.
- Some designs use dual and diverse microprocessor-based control units to satisfy ANSI and IEC standards for PES-based safety systems.
- Output stages of the control unit are configured as for a safety-monitoring relay with dual relay outputs 'K1 and K2' suitable for use with a safety-related control interlock. Additional safety relays may or may not be required according to product selected.

Performance figures for light beams show ranges from 0.5 to 70 m. Small light curtains do not normally require long ranges but even these can operate over 5 m. The main limitation is likely to be found in the alignment and stability of the mountings. Trip response times are typically in the range 15–100 ms.

The majority of AOPDS are produced as light curtains. Let us now consider how the single-beam principle extends to light curtains.

6.6.6   Principles of light curtains

Safety light curtains consist of a strip or line of emitter and receiver pairs that create a multi-beam barrier of infrared light in front of, or around, a hazardous area like robotic cells, loading machines, palletizes, etc. Each emitter/receiver pair operates as described above as a discrete pair with its modulated signal and detection level. Each beam set is operated by the common control unit which now has to serve each pair by multiplexing its input and output signals into a common output stage.

Figure 6.19 shows a circular object that has to be detected by the light beams. The term 'minimum object sensitivity' means the smallest diameter object that will be detected by the light curtain (also known in specifications as the 'resolution'). This effectively dictates the beam spacing or pitch for any particular light curtain. For the principle of redundant channels to be applied for safety this means that the minimum object will have to be detected by two adjacent channels at the same time. The beams therefore have to be aligned very carefully and the detection thresholds have to be set to trip at less than full coverage of the beam.

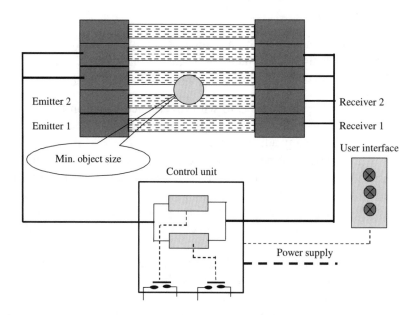

Figure 6.19
Light curtain consists of multiple emitter/receiver sets with common control unit

Standard EN 294 requires resolution of 14 mm for finger detection and 30 mm for hand detection. Hence finger/hand protection by light curtain requires closely pitched beams. These types are used in the 'point of presence' applications where the operator has to reach into the working area to place components or make adjustments (see Figure 6.20).

6.6.7   Safety features of light curtains

In addition to the basic features we listed for the emitter/receiver pair the light curtain assemblies present some additional challenges for the assurance of safety:

- The control unit has to test each emitter/receiver pair for alignment and response.
- There is a risk of divergent beams sending to more than one receiver.

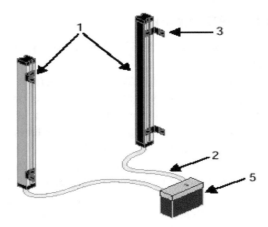

Figure 6.20
Typical opto-safety curtain equipment

- There is the risk of beams being crossed (see Figure 6.21).
- There is the risk of reflections from an object in the field interfering with a receiver.

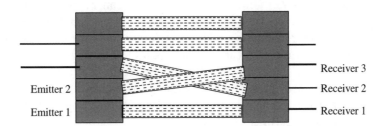

Figure 6.21
Crossed beam error in light curtain

Whilst different manufacturers may tackle the testing and alignment problems differently the general approach is as follows:

- Each beam is given a test pulse in sequence to see that the correct receiver responds to its emitter. This proves that the connection sequence is correct and that each beam is aligned with its correct receiver.
- As each beam is pulsed the other receivers are checked to verify that none of them are switched. This proves that there is no cross-talk due to divergent beams or crossed beams.
- During setting up of the sender/receiver sets the control unit displays the channel number under test and shows the receiver states for all channels. Adjustment proceeds until all channels are receiving on the correct beam. The test sequence then demonstrates a sequential order of switching.
- Test objects are used to verify resolution.

In the workshop we hope to run some demonstration material from manufacturers illustrating some features of the controllers used in light curtains.

6.6.8   Types of AOPD guarding

To conclude this section on AOPDs here are some brief notes on the three main types of optical guarding.

Point of operation guarding   Protects an operator working interactively and close to a machine that is hazardous, typically a small press or assembly station. This needs finger and hand protection at the hazardous point with 14 or 30 mm resolution. Light curtain must be sized to prevent reaching over or under the screen. Requires calculation of safety distance based on human approach speeds, reach distances and stopping time of machine.

Access guarding   Keeps an operator out of reach of hazard points. Hence the screen must be placed at distance beyond reach of an arm or leg plus the distance traveled by the person whilst the machine is stopping. Requires 14 or 30 mm resolution and calculation of safety distance. Can use horizontal or vertical configuration of the screen. Vertical configuration requires additional protection in case a person gets into the space between machine and screen. In the US this second-screen addition is known as 'cascading' or 'host/guest' configuration.

Area or perimeter guarding   Perimeter control light curtains are specified to detect persons crossing a barrier line or gateway into a hazardous area. Hence the beam spacing can be relaxed to a much larger minimum object size. One method is to use two or three small light screens as shown here. Another approach is to use a continuous light screen array but with wider spacing of the beams. Sizes for these applications are chosen to ensure a person is detected. Typically single- or dual-beam units will be mounted at 300, 700 and 1100 mm above ground. Mirrors can be used to include sides but alignment problems will increase. For example, an assembly robot area as shown in Figure 6.22.

Area guarding requires additional safety back-up such as safety controls so that an operator can see that the area is clear before restarting the machine.

In Chapter 7 we shall examine the powerful application features of light curtains that can be obtained through the facilities provided by the more sophisticated control units. These are the features known as muting or blanking and single-break or double-break operating modes (in US PDSI).

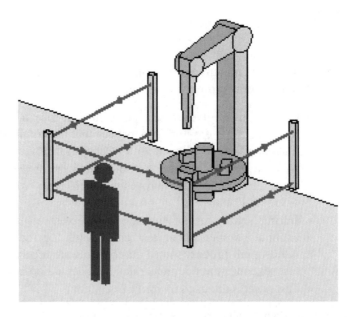

Figure 6.22
Area or perimeter guarding by AOPD

6.6.9   Laser scan sensors for area guarding

These devices employ a laser beam to generate sufficient diffuse reflection for a receiver at the source point to measure the time-of-flight of the light beam. A narrow beam is pulsed with a divergence of typically 1° and then switched by a rotating mirror to the next 1° position for the next pulse. The scan rotates for 180° before sweeping again.

The distance-measuring electronics thereby generates a profile of the field in front of the scanner. Three zones are recognized:

1. Safe zone*:*   No action
2. Warning zone*:*   System can sound a warning alarm
3. Trip zone*:*   Machine will stop.

These devices are well suited to unmanned guided vehicles and are also used for area protection of large machine tools where large pressings or other objects will be moving.

An excellent illustration of laser scan operations is available in the video clip demonstration available from Sick Safety Systems for the 'PLS Area Guarding System'. We expect to show this example in our workshop session. The PLS, for example, has a 4 m radius for its protection zone and a 15 m radius for its warning zone. Object resolution is claimed to be 70 mm. Existing static objects are mapped into the processor memory during setting up which is done through a PC connection.

6.7   Control devices for safety

We saw in Section 6.2 that control devices work by requiring the operator of a machine to stand outside of the danger zone. Hold to run control logic ensures the operator must stay at the safe location to keep the machine running. These are relatively simple and cheap devices, and have been widely used in mass production plants where frequent operator actions are required.

6.7.1   Two-hand controls

This is a device, which requires both hands to operate the machine controls as can be seen in Figure 6.23, and should be installed in accordance with the following:

- The controls shall be separated and protected to prevent spanning with one hand only.
- It shall not be possible to set the dangerous parts in motion unless the controls are operated within say 0.5 s of each other.
- If one or both controls are released, movement of the dangerous parts shall be arrested immediately.
- The controls should be positioned at such a distance from the danger point that, on releasing the controls, it is not possible for the operator to reach the danger point before the motion of the dangerous parts has been arrested.
- Rear and side guarding of the machine are required.

This method was originally not favored because it may be possible with some two-handed controls to tape down one control thus making one-handed operation possible. Even when operated properly, two-handed controls do not protect people other than the machine operator. However for many moderate risk situations these devices offer an economic solution and the technology of the pushbuttons and tamper-proofing has greatly improved.

Figure 6.23
Two-hand hold-to-run control at a workstation

Two-hand controls are made as complete sub-assemblies and a number of safety relay modules are available that allow simple installation to safety category standards. The monitoring units provide self-testing and ensure that the time interval between operations is enforced. The basic performance requirements for two-hand controls and 'hold-to-run' controls are set out in EN 60204 paragraphs 9.2.5.6 and 7.

6.7.2   Hazards of pushbuttons

One of the problems with control devices that use pushbuttons in hold-to-run controls is that they can cause operator strain and repetitive motion injuries such as Carpal tunnel syndrome and tendonitis. To improve this situation manufacturers have developed pushbuttons that require only the presence of a hand or palm or have a very light touch. For example, Rockwell makes a series of 'Zero-Force™ Touch Buttons' using field effect sensing (see Figure 6.24). Another version from Banner uses fiber-optic supplied light beams across a recess in the hand control. These devices are built into two-hand control modules that are supplied with monitoring and control safety relay units to provide a simple safety category rated interface to the machine start/run circuits.

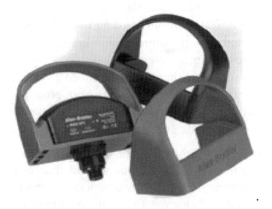

Figure 6.24
Zero-Force™ Touch Buttons

6.7.3 E-stop devices

E-stop devices are not included in the descriptions set down in EN 292-1 as safety devices. However stopping controls are an essential part of the electrical control system for any machine and their requirements fall under the standards for electrical safety of machines. Hence we can find specific details of the stop devices in EN 60204-1 under clause 9: Control circuits and control functions.

The switch is the device that initiates the E-stop. It must sustain this signal until disengaged by the appropriate action. EN 418 is the consultative document for emergency stopping, explaining the differences between the design of a normal stop and an E-stop. It defines the safety requirements of the device as,

> *Having the principle of positive actuation to achieve contact separation that is not dependent on springs or any action on the actuator which generates the signal for an emergency stop must result in a latching of that actuator. The resetting of the actuator shall be only by a manual action.*

The E-stop switch devices shall be of the self-latching type and shall have positive (or direct) opening operation (see IEC 60947-5-1). The actuator may take different forms, depending on the application in which it is being used, for example:

- Mushroom-headed buttons – a pushbutton-operated switch (see Figure 6.25)
- Bars
- Levers
- Kick-plates
- Pressure-sensitive cables
- A pull-cord-operated switch

Figure 6.25
E-stop pushbutton and safety switch

- A pedal-operated switch without a mechanical guard
- Actuators of E-stop devices shall be colored red. The background immediately around the actuator shall be colored yellow. Where used, the background color

must be yellow. In addition to the requirements for stop, the E-stop function has the following requirements:

- It shall override all other functions and operations in all modes
- Power to the machine actuators that can cause a hazardous condition(s) shall be removed as quickly as possible without creating other hazards (e.g. by the provision of mechanical means of stopping requiring no external power, by reverse current braking for a category 1 stop)
- Reset shall not initiate a restart.

The E-stop shall function either as a category 0 stop or as a category 1 stop. The choice of the category of the E-stop shall be determined by the risk assessment of the machine.

Where a category 0 stop is used for the E-stop function, it shall have only hard-wired electromechanical components. In addition, its operation shall not depend on electronic logic (hardware or software) or on the transmission of commands over a communications network or link.

Where a category 1 stop is used for the E-stop function, final removal of power to the machine actuators shall be ensured and carried out by means of electromechanical components.

Devices for E-stop shall be readily accessible. Safety/stop switches must be located where an operator can easily reach them. Poorly located switches may encourage dangerous practices such as reaching across moving parts, a failure to shutdown machinery when a problem occurs or situations where a machine can be started by one worker while another is in a dangerous location (for example, cleaning a bin).

E-stop devices shall be located at each operator control station and at other locations where the initiation of an E-stop can be required.

Where several E-stop devices are provided in a circuit, it shall not be possible to restore that circuit until all E-stop devices that have been operated have been reset.

Because the E-stop device is not likely to be operated frequently, it is recommended that its function be checked (with the guard closed) on a regular basis (start of shift or daily) to enable the safety relays to detect single faults.

6.8   Safety networks and sensors

We have already noted that safety PLCs are an attractive alternative to using a large number of safety relay modules when a safety system is complex. When the architecture of a safety system becomes changed to a PLC basis there are increasing opportunities and needs for sensors to be more efficiently interfaced. Sensors can probably be divided into two classes from a networking perspective.

1. Simple sensors such as limit switches and control switches (including E-stops) must provide fail-safe input data to a safety PLC that will serve to perform the machine trips and interlock, and also serves a powerful visual interface to the operator and the maintenance technician.
2. Complex sensors such as light curtains require their own control units for local safety management but need to pass the basic safety status signals to the PLC. They also stand to benefit from remote monitoring and configuration by network connected PCs.

From these two models we can see how networks can serve sensors (see Figure 6.26).

In the figure the complex sensors such as light curtains each have direct network interfaces. Their control units are able to carry out local management of the sensor system including all diagnostics and fault detection whilst providing all essential data to the host PLC or to other peer devices via the network. Additional fault-detection logic can then be programed in the PLC.

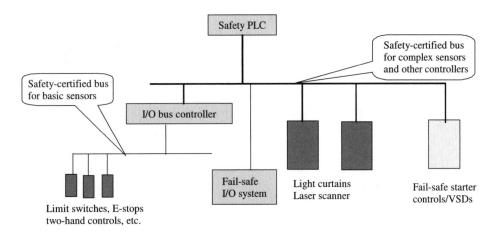

Figure 6.26
Typical requirements for safety sensor networking

For simple devices such as E-stops and limit switches a more attractive approach is to use programmable systems to carry out all the proving and diagnostic functions that would normally be done by a monitoring relay module at each sensor. This requires that application software replaces the logic performed by the safety relays and that a secure fail-safe interface is engineered for the contact status signals of each sensor.

Originally the use of bus systems for safety was not considered to be acceptable due to the risk of undefined failure modes. These objections have been largely overcome due to the development of intensive diagnostic routines for the bus systems and internally in the bus interface electronics at each node. Hence sensors can now be provided with fail-safe bus driver connections and can send and receive data over a safety-certified network. All network functions are built and tested in accordance with IEC 61508 standards for functional safety of programmable electronics.

Example: ASI bus   One example of a low-level sensor network for safety devices is the ASI bus system. This allows a wide range of simple devices to be interfaced to a bus and provides a bus safety-certified controller interface to a high-level field bus system such as PROFIBUS.

Example: PROFIsafe   PROFIBUS is widely used as main automation network protocol with thousands of electronic control devices being available for interconnection. More recently the safety-specific version of this protocol was certified by TUV for use up to SIL–3 class, which is approximately equivalent to category 4 in EN 954. This version is known as 'PROFIsafe'. A particular attraction claimed for PROFIsafe is that it is intended to operate in a shared mode with PROFIBUS on the same network. Hence safety devices and non-safety machine control systems can be built into one integrated environment from a physical aspect but remain completely independent from a functional viewpoint.

Example: SafetyBus-p   PILZ Automation manufactures the SafetyBus-p automation network for the networking of safety-related control systems. In particular the SafetyBus-p supports their PSS range of safety PLC systems.

This design allows sensors to be interfaced to a safety network consisting of many I/O subsystems and one or more safety PLCs as required to manage the network and perform the safety functions. Sensors and tripping actuators can be arranged in groups around the

I/O modules making it easier to take out a section of the network for maintenance without disabling the automation scheme. For more details of the networked safety systems described above, please see Chapter 8.

6.8.1 The E-stop controversy

It should be noted that there is a possible conflict with existing safety standards in cases where a trip function such as E-stop is connected via a network and may even include a safety PLC operating an output signal to the final trip line. The requirements of IEC 60204-1 do not allow this to be done if the stop is classed as category 0:
Extract from para 9.2.5.4.2

Where a category 0 stop is used for the emergency stop function, it shall have only hard-wired electromechanical components. In addition, its operation shall not depend on electronic logic (hardware or software) or on the transmission of commands over a communications network or link.

This prohibition would not seem to apply to category 1 or 2 E-stops as defined by IEC 60204-1, which constitute the majority of stops used automation applications. There is also an ongoing debate on the possibility of revising the category 0 requirements in the light of IEC 61508 certifications. Specialist advice from authorities such as BG or TUV should be sought for more clarity on this issue.

6.8.2 Conclusion: advantages of networking for sensor applications

The advantages of networking systems for safety sensors include:

- Simplified cabling arrangement in the machine installation.
- Eliminates complex interconnection of safety modules to generate the safety monitoring and control functions.
- Safety functions are all configurable and secure in protected software.
- Setting up and fault location becomes simplified by using diagnostic software combined with shared visual display interfaces.

Safety networking is expected to grow rapidly now that certification procedures have been established. Integration of sensors will provide for improved design and upkeep of safety systems.

6.9 Conclusions

In concluding this overview of protection devices and sensors it should be clear that the subject is wide ranging and merits a lot more study by those involved in developing safety systems for machines.

We have attempted to classify the types of protection and identified commonly used sensors. A much greater level of information is available as needed from the vendors of safety devices but the general principles we have covered here are likely to apply and be seen in most of the products available.

Sensors will always be the critical performance items in safety systems. The sensors available meet very high standards of performance and safety integrity supported by third-party certifications such as BG and TUV.

Users have a wide choice of products developed by vendors with long experience of machinery safety applications. The challenge remains to choose the most appropriate solution to each problem.

7

Application guidelines for protection devices

7.1 Introduction

This module will focus how to select and apply the available protection devices to the safety problem. It assumes that we have been introduced to the guards and protection methods available to us as outlined in Chapter 6. In a machinery-building project we have reached the stage where we know about the regulations and the applicable standards, we know about the range of devices available for applications. How do we decide on the best solution for any given problem?

The real problem for the design team is at the conceptual design stage where a realistic solution must be proposed for costing and implementation. To do this we need to have some basic knowledge of what are the good points and bad points of each type of solution and where it is appropriate to use each type. We also need to know some of the essential ground rules for using the chosen method.

In particular this chapter outlines some of the well-established techniques for the application of light curtains to safety duties. These devices are widely applied in automation and in high production rate applications, and it is essential that the designer should recognize the main features that contribute to their success. We can provide only limited coverage of this subject in our workshop but the manufacturers of light curtain systems provide extensive application notes and case studies, which should be consulted for detailed evaluation of any particular application.

7.1.1 Scope and objective

Subjects covered in this chapter include:

- Selection factors for the protection method
- Comparison of physical guarding with other safety methods
- Application of hold-to-run and two-hand controls
- Opto-electronic protection methods using light curtains
- Calculation of safety distances
- Techniques of muting, blanking, single- and double-break operating modes.

The objectives are:

- To be aware of some of the potential solutions available
- To know the factors affecting the choice
- To know the critical features relevant to each type of solution.

7.2   Choosing protection methods

The risk assessment process will have identified the nature of the hazard we are dealing with for any particular application. From that point on the consideration will probably be what type of protection shall we provide? We begin by recalling the main features of protection devices we saw in Chapter 6 and as summarized in Figure 7.1.

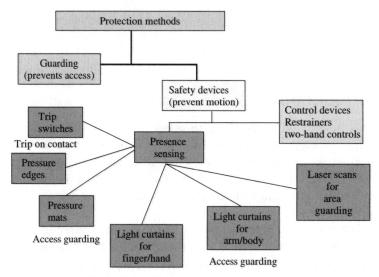

Figure 7.1
The family of protection methods

7.2.1   Physical guarding

Physical guarding prevents access to the danger zone but it brings with it restrictions that can range from inconvenient to unacceptable.

- Fixed guards will do the job when access to the parts is not normally required.
- Interlocking guards allow the guard to be moved out of the way more conveniently and the interlock stops the machine when the guard is open. These are the most commonly seen devices for providing maintenance access to moving parts.
- Key lock systems are available for managing the opening of guards that give access to large machines or assembly lines.
- Control guards speed up the ability to restart the machine after access; hence they can be an integral part of a production process.
- All physical guards can provide protection against objects or hostile conditions and they may be essential for this reason.

7.2.2 Safeguarding devices

This covers the range of devices (other then guards), which can be used to eliminate or reduce risk.

- Preventing machine movement when a person enters the danger zone (either by preventing the start or by stopping the motion). These use presence detection devices combined with an interlock or trip on the machine drives. These methods are widely used where frequent access to the hazardous area is an essential part of the production process (e.g. light curtains for point of operation guarding, point of presence detection, perimeter guarding by light beams).
- Restraining the movement of the machine either by physical means or by a control mode (e.g. slow speed or inching mode controls). These features are of value for the machine setting and specialized operations and are often an integral part of the design.
- Keeping persons outside the danger zone by means of a control device. This requires the person to be out of the danger zone in order to operate the machine (e.g. safety mats, hold-to-run controls and two-hand controls).

7.2.3 E-stop devices

- Allowing a person to command a stop. This is an essential safety function of last resort for all machinery and is mandated by the regulations and standards.

7.2.4 Selection factors

We have already seen examples of devices that fall into these broad categories. The issue now is what is the best device to use in an application? How to proceed?

Our approach here will be to look at the main factors in applying the different types of protection. Very often the choice of protection method will involve a balance between:

- Achieving adequate safety
- What is the most cost-effective or affordable solution?
- Which solution offers highest productivity?
- Which solution is acceptable to and maintainable by the end user?

To balance these factors the designer will need to understand the application rules for each type of protection method. For example, a light curtain solution may look very attractive but the cost may be an issue. Getting the price right depends on how complex the application becomes when all requirements are considered, such as the number of light beams, the spacing and the amount of specialized configuring needed to deal with production objects that will always be interrupting the beams.

We need first to identify some key features and some essential design characteristics to find the strengths and weaknesses of each approach. We can start with basic guarding devices and move toward more sophisticated solutions as we find out the pros and cons of each method.

7.3 Guarding devices

Guarding devices are used for what is known in US regulations as 'point of operation guarding'.

Point of operation guarding means: Machine guards or safety devices such as light curtains that are designed to protect personnel from hazardous machine motion. The guarding therefore has to be applied to surround the point where the hazardous motion is taking place. This distinguishes it from area guarding.

There are two basic options:

1. *Option 1:*   Guards have to be arranged to physically prevent persons or parts of their bodies getting into the danger area.
2. *Option 2:*   Presence detecting devices have to be arranged to prevent the machine making dangerous moves when the person is within a possible zone of danger.

7.3.1   Selection factors for guards

Fixed guards and removable guards with monitoring switches will provide very good protection for those areas where there is no requirement for regular access by the production operators.

Benefits   Simple, secure, protect against flying debris, low cost for small sizes.

Disadvantages   Expensive for large areas. Difficult to apply where automatic feeding and removal is required. Introduces lost time factors in the production cycle.

Since the area being guarded is very often the same point that has to be accessed frequently for the insertion and removable of parts or workpieces we immediately encounter the basic problem of keeping the guard in place at the right time. This applies equally to manually or automated loading of machines. The solution options to this problem are either to automate the guard or use a 'point of presence sensing system'.

Movable and automated guards

This option is widely used and brings with it several factors that form the basis of movable guarding systems.

Guarding developments

Method	Devices	Design Factors
Machines must not start until the guard is in place	Position-sensing switches	Must be used with separate start controls
Machines must not start until the guard is closed and locked	Position-sensing switches + lockable handle with sensing contact	
Machines must stop inside a safety time when a guard is opened	Position-sensing switches. Stopping time monitor	Depends on stopping time being assured. May require monitoring of stop times
Guard must not be opened until the machine has stopped or is in a safe state	Lockable guard with contact on the lock to trip the machine or; Solenoid lock sequenced to lock on start-up and release on zero motion	Depends on time and distance Requires logic control and timer or motion detector. Timer is unsafe if rundown time is variable

Notes:

1. Sensing switches are available, as we have seen, with special features to ensure high reliability for safety and to defeat attempts at bypassing.
2. If a guard can be opened whilst the machine is running the application is unlikely to be safe unless the machine can be stopped before a person can get a hand into the danger zone. Hence the stopping time of the machine is a critical factor. Where this is an unpredictable value the guard will need to be locked by a solenoid under control of the safety system.
3. The release of safety lock by the control system will be controlled by a time delay relay operating after the machine and the stop signal has been received from the final control contactor. A safer option will be to use a stopped or minimum speed detector. Safety modules are available for all these functions, as we have noted in Chapter 4.
4. The release of the safety lock may be conditional on the machine part being parked in a safe position, again a job for a limit switch.
5. Movement of the guard – When guards are to be opened and shut as part of the production cycle there will be a strong case for automatic actuation of the guard. The speed of operation will benefit production rates but the guard movement must not create an additional hazard in itself. It may be necessary to provide edge detectors to protect the operator from the guard!

In practice the application of a guard must take all these factors into account. Many of the type C standards include specific requirements for guards that are based on the above principles. The challenge in each application is to meet the safety requirements whilst avoiding unnecessary loss of production time.

7.4 Point of operation devices

These are the types of devices that prevent the operator from getting into the danger zone of the machine at the time of the dangerous operation. These include the guarding systems we have already discussed as well as:

- Remote controls such as two-hand controls
- Restraining devices to force the operator's hands or body out of the way of harm
- 'Point of presence detectors' or presence sensing devices.

7.4.1 Remote controls

We may decide that it is sufficient to force the operator(s) to step away from a machine by placing the start controls out of harms way. If we use a 'Hold-to-run' control the operator has to keep his finger on the start/run button all the time. This is easy to engineer but it is also easy to defeat by finding ways of jamming the switch. We saw that tamper proof switches are available to deal with this.

7.4.2 Two-hand controls

For many simple applications the use of two-hand control devices provides a simple protection function that allows the machine to be operated without the inconvenience and lost time of a mechanical guard.

In some machines we want to have the operator close to the point of operation but want to ensure that hands are kept away at the time of operation. The two-hand control device was designed for this purpose. It also protects against accidental starting by pressing a wrong button.

Two-hand control devices were described in Chapter 6. The key feature of these controls is to get the position right. The OHSA standard for presses, for example, requires:

> *Application of both of the operator's* hands to machine operating controls and locating such controls at such a safety distance from the point of operation that the slide completes the downward travel or stops before the operator can reach into the point of operation with his hands.

The OHSA standard proceeds to specify the typical concurrent action requirements and specifies the safety distance formula:

> *(vii) The two-hand control device shall* protect the operator as specified in paragraph (c)(3)(i)(e) of this section.
>
> (a) When used in press operations requiring more than one operator, separate two-hand controls shall be provided for each operator, and shall be designed to require concurrent application of all operators' controls to activate the slide. The removal of a hand from any control button shall cause the slide to stop.
> (b) Each two-hand control shall meet the construction requirements of paragraph (b)(7)(v) of this section.
> (c) The safety distance (D_s) between each two-hand control device and the point of operation shall be greater than the distance determined by the following formula:
>   - $D_s = 63$ inches/second $\times T_s$ where:
>   - D_s = minimum safety distance (inches); 63 inches/second = hand speed constant; and
>   - T_s = stopping time of the press measured at approximately 9 deg. position of crankshaft rotation (seconds).
> (d) Two-hand controls shall be fixed in position so that only a supervisor or safety engineer is capable of relocating the controls.

Benefits:

- Low-cost solution uses modular two-hand control devices available from several vendors.
- Devices are supported by modular two-hand control relays with all timing functions provided and with facilities for adjustment.
- Control device and application well defined by standards, e.g. EN 574.
- Two-hand control can be particularly useful in applications such as teach mode pendants and inching (or jogging) controls because it can give enhanced levels of protection when used in conjunction with other protective devices.

Disadvantages:

- Lost productivity if other tasks could be performed at the same time. If used for the start of a long or continuous running operation this method would be a complete waste of time.
- For frequent cycles there will be lost time whilst the operator walks to the two-hand control station.

- Not safe if any other person can get into the danger area at the same time of operation.
- If more than one operator is required a second control unit must be used. Control station designs require special features such as covers to defeat bypassing. The control unit must be fixed in its safe location.

Using two-hand control stations with back up One method of improving safety levels with a two-hand controller is to provide additional safety measures such as a safety mat in combination with a two-hand control device. Then if any other person should stray into the area or if the operator cannot see clearly from his workstation the safety mat will still trip the machine.

7.4.3   Mechanical pull out devices

These devices have been used historically in days of purely mechanical protection systems.

1. *Pull out devices*:   Attachments to the operator's hands that pull them out of range of the press dies before the press operates. These are mechanical devices linked to the press machine that pull back the operator's hands as the press begins its stoke. This type of device and the following variant are not likely to be found in modern production plants but provide a basic 'low-tech' solution.
2. *Restraining devices:*   Also mechanical devices that attach to the operator and prevent his hands from reaching the danger area. *Effectively he is tied up!*

7.4.4   Point of presence detectors

This term covers all the devices that can be used to detect a person or part of a person approaching or actually in the danger area. These include:

- Pressure-sensitive mats
- Pressure-sensitive edge detectors
- Opto-electronic devices such as light curtains and light beams.

Presence detectors depend on the combination of the detection of a presence and the rapid stopping of the machine motion. This leads to the *provision of a 'safety distance'* to ensure that a machine is in a safe condition before anyone can get any part of their body into the danger zone.

The OHSA standard introduced a formula for the safety distance between the point of detection and the danger area of the machine calculated from a hand movement speed of 63 in./s multiplied by the stopping time of the press after initiation of a stroke. We quote from the standard:

(iii) A presence sensing point of operation device shall protect the operator as provided in paragraph (c)(3)(i)(a) of this section, and shall be interlocked into the control circuit to prevent or stop slide motion if the operator's hand or other part of his body is within the sensing field of the device during the down stroke of the press slide.

> *(a) The device may not be used on machines using full revolution clutches.*
> *(b) The device may not be used as a tripping means to initiate slide motion, except when used in total conformance with paragraph (h) of this section.*
> *(c) The device shall be constructed so that a failure within the system does not prevent the normal stopping action from being applied to the press when required, but does prevent the initiation of a successive stroke until the failure is corrected. The failure shall be indicated by the system.*

(d) *Muting (by passing of the protective function) of such device, during the upstroke of the press slide, is permitted for the purpose of parts ejection, circuit checking, and feeding.*

(e) *The safety distance (D_s) from the sensing field to the point of operation shall be greater than the distance determined by the following formula:*

$$D_s = 63 \text{ inches/second} \times T_s$$

where:

D_s = minimum safety distance (inches); 63 inches/second = hand speed constant; and

T_s = stopping time of the press measured at approximately 90 degree position of crankshaft rotation (seconds).

(f) *Guards shall be used to protect all areas of entry to the point of operation not protected by the presence sensing device.*

This was the beginning of a more sophisticated set of safety distance rules that have evolved into the standards now available from ANSI and from the type B standard EN 999. We shall look more closely at distance calculations in the next section. For the moment we shall complete our review of the pros and cons of different presence sensing methods.

7.4.5   Pressure-sensitive safety mats

These devices are used to guard a floor area around a machine. A matrix of inter-connected mats is laid around the hazard area and any pressure (e.g. an operator's footstep) will cause the mat controller unit to send a stop signal to the guarded machine.

Pressure-sensitive mats are often used within an enclosed area containing several machines, e.g. flexible manufacturing or robotics cells. When access may be required into the cell (for setting or robot 'teaching', for example) they prevent dangerous motion if the operator strays from the safe area. The size and positioning of the mats should be calculated using the formulae from the standard EN 999 'The positioning of protective equipment in respect of approach speeds of parts of the human body'.

Benefits:

- Adaptable and reasonably low-tech solution.
- Suitable for area protection of continuous working plant.
- Can be customized to many applications by the end user as the shop floor layouts are developing.
- Leaves the operator's hands free to work at the control panel or collect new parts.
- Useful for robot teaching tasks where technician has to work with a robot.
- Control device and application well defined by standards.

Disadvantages:

- Possible to defeat unless carefully designed and well policed.
- Requires all detection to be linked to the position of the person's feet. Hence they cannot serve for point of operation guarding where the operator must be allowed to stand close to the working area.
- Safety mats risk wear and tear, and attack by oils and chemicals.

7.4.6   Edge-sensitive devices

Pressure-sensitive edge strips are used for applications such as power-operated doors, guided moving vehicles and moving machinery beds. They can provide a continuous line of touch sensing along surfaces. If the moving part strikes the operator (or vice versa) the flexible sensitive edge is depressed and will send a stop signal to the power source. Sensitive edges can also be used to guard machinery where there is a risk of operator entanglement.

Such devices will allow access where required, without the constraints of mechanical interlocked guards. For example,

- Contact sensing bumpers can be used on safe edges on numerous machine applications or as bumpers on automatic guided vehicles (AGVs).
- If an operator becomes caught up and dragged by the machine he will be pulled onto the sensitive edge thereby tripping its switching action.

They rely entirely on their ability to both sense and switch for the provision of safety, therefore they will be connected to typical guard switch or E-stop monitoring relay. The monitor will need to be arranged to check for circuit continuity in designs where the sensor contact is to be closed when the pressure applied to strip.

In general pressure-sensitive edges are only suitable on machinery, which can be stopped quickly enough to avoid harm to the person in contact. It is necessary that the time taken for the motion to stop be less than that required for the operator to reach the hazard after tripping the device.

Benefits:

- Adaptable low-cost solution for minor injury situations
- Well proven in applications such as subway train doors.

Disadvantages:

- Not suitable for point of operation guarding
- Not suitable for severe injury risks.

7.4.7   Protection methods using light curtains

The problem of lost time between machine cycles in machines such as presses presents a challenge to safety system designers to find the most efficient technique to ensure safety without sacrificing productivity. The best solution very often lies in the technology of light curtains or opto-electronic guarding as we saw in Chapter 6. Here we summarize the main types of protection offered by light curtains or scanners:

- *Point of operation guarding:* Finger and hand protection for close working at a machine. This allows an operator to reach into a hazardous working point such as the tool of a press at a time when it is safe by virtue of the state of the machine. At any other time the machine will be stopped whenever the light curtain is interrupted. The control point for starting and stopping the machine must of course be positioned outside of the hazardous zone.
- *Area guarding* is very similar to point of operation guarding but a hazardous area is defined by the designers rather than a targeted point in the machine. The light curtain must be carefully laid out to ensure that there are no pockets of space between the light curtain and the machine that a person could occupy and be undetected. This method also requires that the machine controls be placed

out of range of the hazardous area. Hence safety is assured by the following principles:

- The machine trips out when someone enters the area.
- The safety distance is satisfied by the position of first detection.
- The presence of a person or a limb in the hazardous area is detected at all times.
- The trip cannot be reset until the hazardous area is clear.
- The machine cannot be restarted until the trip is reset and the controls are operated.
- The reset and controls are located outside the hazardous area but in line of sight of the hazardous area.
- The trip reset can be automatic or manual. Restart of the machine is by the operator except in the special case of PSDI operation as described later.

- *Access guarding or perimeter guarding:* The light curtain detects a limb or body of a person gaining access to a hazardous area but the area is too large to be mapped by the light curtain. By the use of mirrors the light beams can be diverted round corners to enclose a machine unless rigid walling is available for the sides. The result can be regarded as a kind of 'optical fence'. The safety distance for the perimeter line must be set so that no one can reach the machine before it can be stopped. Access guarding is widely used for putting a protected area around a robotic welding or assembly area since it gives complete freedom of access as soon as it is needed (see Figure 7.2).

Figure 7.2
Perimeter or access guarding

- *Area guarding by scanning:* This technique extends the principle of area guarding to deal with larger areas by scanning at high speed using a laser beam (see Section 6.6.5). Fixed areas are configured to allow for static objects so that the system can detect any other object in the field that should not be there. This system is suggested for large machine tools with some support from perimeter guarding and most likely a point of operation guard for the tool area.

There are great benefits to be had by using light curtains in many safety applications except where a physical shield is needed against falling into the machine or to protect a person from flying debris. Technical developments in the control modules have added flexibility and configuration capabilities that make light curtains adaptable to many automated manufacturing processes. In the next section we shall look at the special features of light curtain systems that make them particularly adaptable to automation and materials handling applications.

At the same time it should be noted that the light curtain installations must be carefully planned and evaluated for performance factors such as object sensitivity and safety distances. The cost of an installation can increase substantially as detailed evaluation arrives at the need for more beams or additional features.

Benefits:

- Products available for a wide range of applications
- Self-testing and fail-safe characteristics up to category 4
- Supports high productivity by allowing unrestricted movement of persons and hands-free operation
- Configuration tools allow flexibility in response with options such as muting and blanking. Adaptable quickly to product changes in the machine
- Suitable for guarding of robot-operating areas
- Suitable for advanced manufacturing units and assembly lines through programing options, e.g. blanking, muting, PSDI (double break initiation)
- Laser scanning versions suitable for automatic guided vehicle applications.

Disadvantages:

- Not suited to low-technology environments
- Relatively high cost
- Specific versions needed for each application
- Requires careful application engineering and validation.

Conclusions on the choices

A logical selection process can quickly narrow down the wide range of products to be used.

- If you require frequent access to the working parts for production the choice is between interlocked/control guards or safety devices.
- Safety devices of all types allow faster interactions with the machine than guards but may sacrifice the level of safety that can be assured.
- Control devices offer a low-cost solution but may not give adequate safety in cases of severe consequence due to the chances of out other persons getting in the way of the machine.
- Light curtains can achieve high safety and high production rates but at a higher initial cost. This has to be offset against possible improved returns.
- For complex automation tasks light curtains may be the only satisfactory solution.

For example, Rockwell Automation makes this comment regarding the merits of light curtains:

Typically, a safety interlock gate is used to help prevent machine motion when an operator enters the hazardous area. Even if it only takes 10 seconds to open and close that gate for each cycle, that time accumulates over the course of a 200-cycle day. If the traditional gates were replaced with light curtains, operators would simply break the infrared barrier when entering hazardous areas and the operation would come to a safe stop. Over time, the light curtain investment would increase productivity and create a positive return.

Examples include automated warehouses with conveyors and guided vehicles. Access control typically uses single- or wide-spaced light beams to detect the entry of a person into the hazardous area and trips machines on entry.

7.5   Application guidance notes for light curtains

At this stage we need to be aware of a number of techniques and some terminologies in opto-electronic safety systems. This will help with judging the suitability and complexity of a device for an application. We can only outline some of the main features in this workshop to serve as an introduction to the many possibilities of the technique. The manufacturers of opto-electronic protection systems all provide very helpful guidance on their products at the detailed level.

Our notes here are based on guidance notes available from manufacturers of opto-electronic protection systems including Sick Safety Systems, Banner Engineering, A.B. Guardmaster. Please note that websites for all these vendors are given in Appendix.

Summary of techniques

The techniques we are going to outline here are:

- Design steps for specifying an optical guarding system (AOPD)
- Calculation of safety distances for point of operation guarding of a machine
- Principles of blanking and muting
- PSDI or single-break/double-break operating mode.

Terminology

Some of the basic terms used for the opto-electronic systems are defined here. Please note that the best reference for the specialist terms is EN 61496-3.

- *AOPD (active opto-electronic device):*   The term for the optical sensor system used in the EN standard: EN 61496. This type B2 standard defines the application requirements for AOPDs and should be used as the primary reference for their application.
- *ESPE (electro-sensitive protection equipment):*   This is a generic term for a complete system of sensor, evaluation unit and output switching stage of any electronic detector system. It comes from EN 61496-1.
- *Minimum object sensitivity:*   Defined as the minimum detectable diameter of a cylinder placed in the field of the light curtain. Obviously for a small object to be certain of detection the beams must be close together. Hence it will become impracticable and expensive to have small object sensitivity in a large light curtain.
- *Muting/blanking:*   This is a term used to describe the blocking of certain beams in the light curtain to allow for objects that may obstruct the beam but do not represent a hazardous intrusion. Extends to floating blanking (USA) or muting (EU) to deal with legitimate moving objects generated by the production process.
- *PSDI (presence sense device initiation):*   A technique that allows machine cycles to be started by detecting the operator as withdraw from the hazardous area. Similarly in Europe it is applied under the name of single/double-break operating mode. The requirements in Europe are defined in EN 61496 (see later in this chapter).

7.5.1   Design steps for specifying an AOPD

The steps for specifying an opto-electronic guard are as follows.

Step 1: Define the zone to be guarded

This is based on the machine's risk assessment, in which access to the danger zone can be specified, i.e. the configuration of the installation, size of the protective field and space available in front of the danger zone. The dimensions of the machine also play a part in determining the screens required to guard the area. This necessitates the need for calculation of safety distance to mark-up the zone to be protected.

Step 2: Define the safety function

Define the safety function to be performed with respect to ergonomic criteria and ergonomic factors (e.g. cyclic insertion of parts or no cyclic entry).

Here one will need to define exactly what is to be detected within the danger zone:

- A finger or hand (required when the operator is near to the hazard). In all cases, the resolution of the AOPD must be less than or equal to 14 mm.
- Arm or body (mainly for perimeter guarding).
- Presence of an operator (especially where the guarded machine is not visible from the control point). This is also suitable for guarding the approach to danger zones, and where vehicles are involved.

Based on above the classification could be,

- *Point of operation:*   For point of operation guarding the hazard will need detection for fingers and hands, so the specification will call for a minimum object sensitivity of 14 or 30 mm as the standard value. The aperture to be covered will define the length of the beams and the height or width of the curtain, hence defining the number of beams.
- *Perimeter or area guarding:*   Will require the AOPD to be of typically a 9 m range, which can achieve an object sensitivity of 20 mm. An 18 m range can achieve 25 mm. Alternatively a scanning laser device can cover a large area by sweeping and detect the presence of an object in its field. Available units have a 4 m protection range with a 180° arc, resolution of 70 mm.
- *Access control:*   Whole body size or robot arm or vehicle one or two single-beam devices or with larger spaces between beams.

The safety function will also have to consider the additional performance requirements that may be needed for practical operation. For example, there may be a need for blanking or muting.

Single light beams will not normally provide suitable protection, as it will usually be possible for a person to reach around the light beam and reach a dangerous part. A satisfactory arrangement can be provided by placing a number of light beams so that there are no gaps in the arrangement of light beams that will allow people to reach through, around, under or over the dangerous parts.

Step 3: Define the safety category

Note that many severe applications have a mandatory category 4 requirement from the relevant type C standard for the machine. The AOPD standard defines safety categories but uses the word 'type'. A 'type 4' safety system is considered to be very similar to a 'category 4'.

Step 4: Determine the types of approach

Here we are going to combine a study of the approach layout with the calculation of safety distance as done earlier while determining the zone of protection. The safety distance is determined first using guidelines based on the standard EN 999 as explained above. Then the positioning of the screen to achieve the safety distance must be decided. There may be a choice of how the approach layout is to be arranged and we would like to find the most reliable and hopefully the most economic arrangement.

7.5.2   Calculation of the safety distance

The safety distance is calculated from the knowledge of the stopping time of the overall system (T) made up from component elements, as well as the resolution of the AOPD which has a minor effect on the penetration of an object before detection. The calculation is adjusted to arrive at the most satisfactory balance between the safety distance (S) of the guard and the achievable stopping time of the machine. Figure 7.3 illustrates the process with parameters shown.

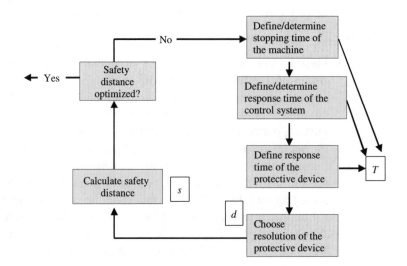

Figure 7.3
Process of optimization of safety distance against stopping time

Following the definition of the area to be protected, the safety distance is defined by the following parameters:

- Stopping time of the machine
- Response time of the control system
- Response time of the electro-sensitive protective equipment (ESPE) (note that the term ESPE implies the complete system response rather than just the optical sensor)
- Additional margins on the safety distance calculated.

The calculation of the safety distance for an ESPE is described in the standard EN 999. If the machine is the subject of a specific standard (e.g. metal presses) or a special technical specification, reference must be made to this document. The general formula from EN 999 is:

$$S = (K \times T) + C$$

where

- S is the minimum distance in millimeters, measured from the danger zone to the detection point, to the detection line, to the detection plane, or to the protective field.
- K is a parameter in millimeters per second derived from data on the approach speeds of the body or parts of the body.
- T is the run-on time of the entire system in seconds.
- C is an additional separation distance in millimeters that defines the penetration into the danger zone prior to the triggering of the safety device.

Based on the resolution of the AOPD, in the calculation the additional margin C must be added to the minimum distance S. The distance C is to be based on performance factors for the light curtain such as the intrusion before the detector will switch. This factor allows for the ability of a person to reach over the detection beam and is therefore a function of the height of the beam. If the minimum distance is too large and not acceptable for ergonomic reasons, it must be established whether it is possible either to reduce the overall stop time for the machine or to select an ESPE with better resolution.

EN 294 defines 'safety distances to prevent danger zones from being reached by the upper limbs'. For example, in the case of vertical approach and an overall stop time of 100 ms, for an AOPD with a resolution of 35 mm, a distance of 368 mm is calculated. Conversely, with a resolution of 14 mm, the calculated distance is only 200 mm.

When applied to presses, for example, the standards EN 692 and EN 693 call for the following values of C as a function of the detection capability of the AOPD.

Resolution d (of AOPD) in mm	Additional Margin C (mm)	Stroke Triggering by AOPD/Cyclic Operation
≤ 14	0	Permitted
>14 to ≤ 20	80	
>20 to ≤ 30	130	
>30 to ≤ 40	240	Not permitted
>40	850	

Each AOPD must be installed such that access to the danger zone without detection by the protective device is impossible.

For all guarding duties the first point to check is that it must not be possible for the operator to get any part of his or her body between the screen and the dangerous area. For many machines this is mandated in the C standard but it is an obvious requirement that must be carefully observed. The object sensitivity dimensions for fingers, hands and limbs have been set down in the standards. The upper and lower heights of the screen are also specified.

Some guidelines on safety distances and layouts are given here based on material published by Sick Safety Systems and by PILZ and using the EN 999 formula. Note that a more detailed method of calculating the C- factor is given in these formulae.

Generally we can distinguish between three types of approach:

1. Perpendicular
2. Angular
3. Parallel.

Perpendicular approach

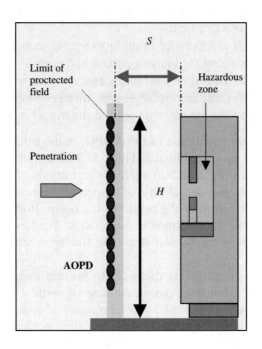

Figure 7.4
Perpendicular approach/entry perpendicular to the plane of the protective field

For the perpendicular approach as shown in Figure 7.4, the following guide formulae quoted by Sick Safety Systems apply for the safety distance *S*:

$\beta = 90° (\pm 5°)$ $d = <40$ mm	$S = 2000$ (mm/s) $\times T + (8 \times (d - 14))$ where $S > 100$ mm where $S > 500$ mm take $S = 1600T + 8(d - 14)$ In this case S cannot be <500 mm	S: minimum distance H:height d: resolution β: angle between plane of detection and direction of penetration T:time
40 mm $< d \leq 70$ mm	$S = 1600T + 850$	Height of lowest beam ≤ 300 mm Height of highest beam >900 mm

$d > 70$ mm
  multi-beam
	$S = 1600T + 850$	No. of recommended beams and
		  heights
		4 300, 600, 900, 1200 mm
		3 300, 700, 1100 mm
		2 400, 900 mm
		1 750 mm

Single beam	$S = 1600T + 1200$	

The height of the highest beam in the light curtain is determined by the height of the danger zone relative to the person at risk. Values for the heights are given in EN 294 'Safety distances to prevent danger zones from being reached by the upper limbs'. For example, with a danger zone height of 1000 mm above ground EN 294 calls for the top beam to be at least 1800 mm above ground.

Angled approach

Formula for calculating the safety distance when using angular approach (see Figure 7.5):

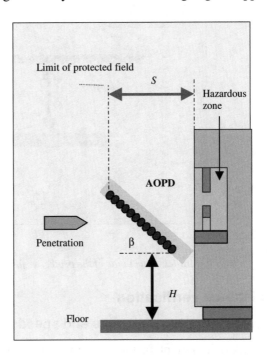

Figure 7.5
Angled approach/entry to the plane of the protective field

$5° < \beta < 85°$	Where $\beta > 30°$, cf. perpendicular approach; where $\beta < 30°$, cf. parallel approach; S then applies to the furthest beam whose height ≤ 1000 mm	$d \leq H/15 + 50$ applies to the lowest beam

Horizontal appproach

Formula for calculating the safety distance when using parallel approach (see Figure 7.6):

$\beta = 0° (\pm 5°)$	$S = 1600T + (1200 - 0.4H)$ where $1200 - 0.4H > 850$ mm	$15 \times (d - 50) \leq H \leq 1000$ mm. where $H > 300$ mm there is a risk of undetected access under the beam to be taken into account for H where $d \leq H/15 + 50$

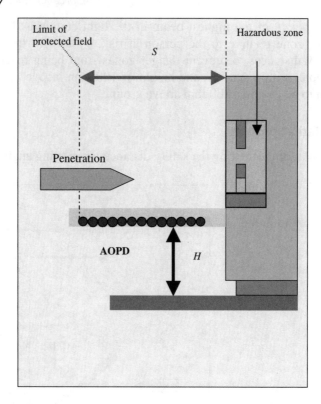

Figure 7.6
Parallel approach/entry parallel to the plane of the protective field

7.5.3 Safety distance verification

Monitoring of stopping distances and speeds

One of the most important factors affecting safety distances is the extent to which the machine stopping time can be assured. The press regulations require that the safety system include the facility to monitor the press brake performance. If the press clutch and brake devices become worn there is a risk that this stopping time will become longer. Hence the safety distances will be invalid. This leads to the requirement that the stopping time shall be monitored and the press shall be tripped out if the time exceeds the original setting by more than a given factor, typically 10%.

Monitoring safety modules are available for these duties and should be included in the safety function design for any machine where the stopping time may be liable to drift out with age or with modification of the machine usage. The following extract from OHSA regulations is relevant to illustrate this point.

Extract from OHSA 1930 section 217:

> ***14) Brake system monitoring.*** *When required by paragraph (c)(5) of this section, the brake monitor shall meet the following requirements:*
>
> > (i) *Be so constructed as to automatically prevent the activation of a successive stroke if the stopping time or braking distance deteriorates to a point where the safety distance being utilized does not meet the requirements set forth in paragraph (c)(3)(iii)(e) or (c)(3)(vii)(c) of this section. The brake monitor used with the Type B gate or movable barrier device shall be installed in a manner to detect slide top-stop overrun beyond the normal limit reasonably established by the employer.*

(ii) *Be installed on a press such that it indicates when the performance of the braking system has deteriorated to the extent described in paragraph (b)(14)(i) of this section; and*

(iii) *Be constructed and installed in a manner to monitor brake system performance on each stroke.*

7.5.4 Conclusions on safety distances

The above examples show that for each application there is some basic design work to be done to find the most cost-effective arrangement of light beams consistent with safety. The most important rules are:

- The safety distance must be calculated using standards and recorded on file for checking.
- The arrangement must not permit a person to be undetected in the space between the light curtain and the hazard.
- Stopping times must be monitored as part of the safety function.

Bearing in mind that light curtain systems are offered in kits with control units and a wide range of sensor sizes it is reasonable to expect that good solution can be found for most applications.

7.5.5 PSDI or single/double-break initiation

A section of the US power press regulations is devoted to the specification of PSDI operations. This operating mode is advantageous if parts are inserted and removed cyclically by hand. In this mode the machine cycle is automatically reinitiated once the protective field becomes clear after interruption once or twice.

The reset device is to be operated under the following conditions:

- On machine start
- On restart if the AOPD is interrupted during a hazard producing movement
- To initiate a restart after a cycle duration of more than 30 s (EN 61496).

More detailed information is to be found, e.g. in EN 692. Nevertheless, it is necessary to check that no hazard for the operator can be generated during the work process. This limits application to small machines on which it is not possible to walk into the danger zone, and on which there are guards against walking behind.

All other sides of the machine must also be safeguarded using suitable means. If this operating mode is used, the resolution of the AOPD must be less than or equal to 30 mm (cf. EN 999, EN 692, prEN 693).

7.5.6 Blanking

Blanking is a general term used in the USA describing the feature within a light screen system that allows it to ignore a certain object within its field. The instruction to ignore a particular beam is programed into the control module under supervisory key or password protection.

- Fixed blanking is for a fixture on the machine or objects permanently in the field. The possible problem with fixed blanking is that it may impair the detection of a finger at that point; hence it has to be carefully worked out by studying the layout of the hazardous area.

- Floating blanking (US term) caters for a known moving object to pass through the beams of the curtain without causing a trip. This is known in EU standards as muting (see below).

7.5.7   Muting of protective devices

The muting or temporary bridging of protective devices raises the problem of an installation's safety. For example, EN 415-4 (Palletizers and depalletizers) relates to packaging machinery on which all operations on the palletized load are carried out entirely and automatically by machine. At the inlet and outlet of the inner chamber (where there is a risk under normal conditions) it is necessary to bridge the AOPD at the moment when the pallet passes through.

On the other hand, it is necessary to detect the entry of persons. Under normal operating conditions, there is a risk at both the entrance and exit of the interior zone. The AOPD must be muted at the moment the pallet passes through, but it must also be possible to detect the presence of an operator. The muting system must therefore be able to discriminate between the pallet and the operator.

The muting conditions defined in EN 415-4 state that:

1. Muting may only occur during the time span of operating cycle when the loaded pallet obstructs access to the hazardous or danger zone.
2. Muting must be performed automatically.
3. Muting must not depend on a single electrical signal.
4. Muting must not depend entirely on software signals.
5. The muting signals, if they occur as part of an invalid combination, must not permit any muting state and ensure that the protective function is retained and the machine locked out.
6. The muting state is deactivated immediately after the pallet has passed through and the protective device is thus effective again.

Please note that excellent detailed illustrations of blanking and muting can be found in the guide manual and information CDs available from the leading optical device suppliers including Sick, Banner and A.B. Guardmaster.

These devices provide temporary bridging in a system by means of automatic differentiation. This means that the relevant device is very easy to understand for the user who does not need to address the wiring for the automatic, temporary bridging.

A light curtain can be used to meet all these requirements. The device incorporates a system of temporary muting by automatic discrimination. Auxiliary light beams are used to detect the approach of a permitted object and the action of the main detection curtain is suspended or muted until the object has cleared the auxiliary beam at the exit stage.

7.6   Conclusions

It should be clear from the brief examination of the protection systems that why the light curtain devices have become so dominant in the more sophisticated machinery applications. Their application requires considerably more training than we have been able to cover here but the rewards in productivity are known to be substantial. The devices have become powerful and relatively easy to use through the growth of microprocessor-based designs. Software and hardware for these systems can now be produced and certified in accordance with the programmable safety systems standard IEC 61508. The potential of these systems will continue to grow.

For simpler applications and for a wide variety of tasks the large range of guard switches, safety gate monitors and new miniature optical devices used for limit switch duties will always be used in large numbers. The selection process depends on a methodical approach by the designer balancing costs against productivity benefits within the bounds of acceptable safety levels.

8

Programmable systems for safety controls

8.1 Introduction

This chapter investigates the following topics:

- How might a PLC be arranged to serve for safety control functions?
- What are the potential benefits to the factory?
- What are the essential performance requirements for a safety PLC and why is a general purpose PLC unacceptable for safety?
- What are the key design features of a safety-certified PLC?
- What types of products are available?
- The development of safety-related field bus systems.

Now that we are aware of some of the basic practices in machinery safety systems we are ready to look at how programmable control systems can provide benefits to users of the more complex types of machinery or to those building up a production line.

The introduction of specially designed and certified safety PLCs during the mid-1990s has been supported by the development of new engineering guidelines and standards for their application. Today the application of safety PLCs in machinery safety is well established and is growing rapidly in scope. The safety PLCs combined with the development of safety-related bus technologies and field-bus-connected safety devices are bringing many changes to conventional safety control practices. We shall try in the next two chapters to introduce some of the most important aspects of the use of PLCs in machinery safety.

8.1.1 Purpose and objectives

This chapter has three related purposes:

1. To show how programmable systems can be arranged to provide safety functions and to identify the potential benefits to be obtained.
2. To explain why safety PLCs are different from general purpose PLCs.
3. To illustrate key design features of some typical safety PLCs.

The objective of these topics is to ensure that workshop participants are able to participate effectively in projects involving PLCs in safety.

Principles of the PLC and its use in safety control functions

We begin by considering the essential characteristics of a PLC and seeing how it fits into the safety-related control models we first saw in Chapter 1.

Figure 8.1 shows the essential parts of any safety-related control system. The hard-wired and modular equipment we have seen earlier in the workshop fits into this model. The element in the center can be electrical/electronic or programmable electronic in nature. If we use a PLC for the logic solver it fits the model exactly.

Elements of a safety-related control system

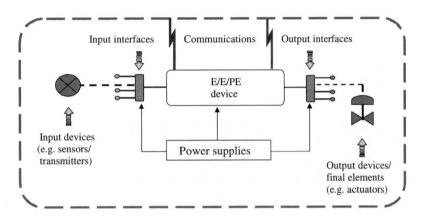

Figure 8.1
Safety-related electrical control system basic diagram

Figure 8.2 uses a PLC for the logic solver element of the safety function and provides the input and output interface through its input and output subsystems.

PLC serving as safety-related controller

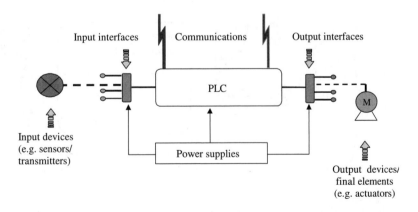

Figure 8.2
PLC version of safety controller

If we place this control system into the context of a machinery control system using the diagram we introduced in Chapter 1 it will look like Figure 8.3.

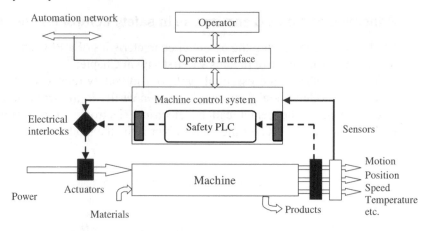

Machine with safety-related electrical parts using a safety PLC

Figure 8.3
PLC serving as logic solver in an SRECS

In Figure 8.3 it's clear that the PLC that is providing the safety function fits into the overall machine control system as a safety-related part. It reads in the safety sensors via its interface modules and outputs commands via the interface to the final control elements (which remain hard-wired and electromechanical). The main difference between this version and the hard-wired safety-related controls is that we would normally plan to share all the safety functions in the same safety PLC.

In Figure 8.4 the PLC provides the logic solver subsystem for several safety functions. It follows that the PLC must then be engineered to meet the highest safety integrity level required for the functions. For example, if one of the functions is a category 4 and the others are category 2, the PLC systems will have to meet category 4.

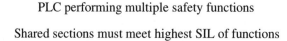

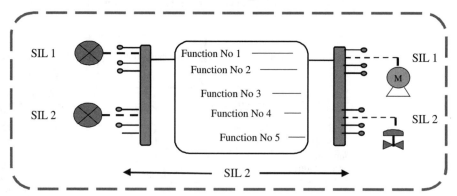

Figure 8.4
PLC serving as logic solver in an SRECS

This means that we cannot just use the same PLC that we may have installed for the basic control functions of the machine unless that PLC can meet all the special safety requirements needed to provide the equivalent of say a category 4 hardware safety function. We need to have a PLC that is designed for the job.

8.1.2 Options for the machinery control system

The options for a complete machinery control system are then:

(a) Use a standard PLC for basic control with a separate safety PLC for the safety functions.
(b) Build all the machinery controls into a safety PLC and apply safety practices to all control functions.
(c) Use an integrated control and safety PLC system.

8.1.3 The option of separate PLCs for control and safety

Option A is a generally workable solution that is widely used and is represented by Figure 8.5.

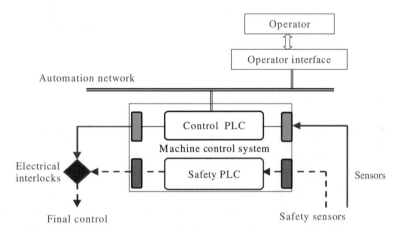

Figure 8.5
Machine with separate PLCs for safety and control

With reference to Figure 8.5:

• The safety-related functions are performed by the safety PLC and there is no danger of interference from the control PLC.
• The safety PLC will have a well-defined set of functions to perform and it can be selected at the right size and cost for the job.
• The basic control PLC may be a large or a small system chosen from the vast range of products available and its specification is not affected by the requirements of the safety system.

8.1.4 Interfacing to the safety PLC

Figure 8.5 does not show how the operator interface is arranged for the safety PLC, but it will soon become apparent that there is a strong case for safe networking of the safety PLC. For the moment we should note the following:

• Hard-wired input and output devices for the essential safety functions, such as E-stops and indicators, can provide the essential operator interface.
• The engineering interface can be by a dedicated programing port or by a secure network connection.
• If the safety PLC has a secure network connection, this can be used to send status information to the basic control system and to the main operator interface.

8.1.5   Using a safety PLC for all controls

Using option B, a safety PLC could be installed for all control duties as shown in Figure 8.6.

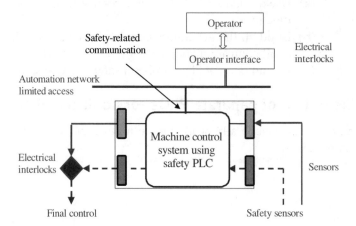

Figure 8.6
Machine with single safety-certified PLC for safety and control

With reference to Figure 8.6, the single safety PLC option:

- The single option means that all control functions will have to be engineered using the safety-related hardware and software in the safety PLC package.
- Communications to operator and automation are subject to safety limitations.
- All safety management precautions apply to all functions in the controller.
- This generally makes it expensive and it limits performance flexibility for the basic control sections.
- This is sometimes a workable solution for a machine where nearly all of its control functions are considered to be safety-related.
- It has the advantage that a single system is used for the machine.

8.1.6   Using an integrated control and safety PLC system

Our third option, C, seeks to achieve the best of both worlds by building an integrated PLC system that offers all the flexibility and power of a control system PLC with a special section designed and certified for safety duties. Effectively this provides for the 'safety-related parts' of a PLC. This arrangement is depicted in Figure 8.7.

With reference to Figure 8.7, the integrated control and safety PLC option:

- Combining the two PLCs means that all control functions can be engineered using a single programing facility. The programing facility permits standard control functions to be engineered with a full range of application function blocks whilst the safety functions are only permitted to use approved safety function blocks.
- I/O subsystems can be provided for both standard and safety functions under a shared bus control system.
- Cost saving is achieved by making best use of standard non-safety-related parts for all standard I/O duties and using safety-related parts only for safety duties.
- All communication facilities are available to both safety and non-safety sections. Secure interfacing is managed within the bus structures of the PLC.

- Communications between safety and non-safety functions are safe and simple within the program structures.
- It has the advantage that a single system is used for the machine.

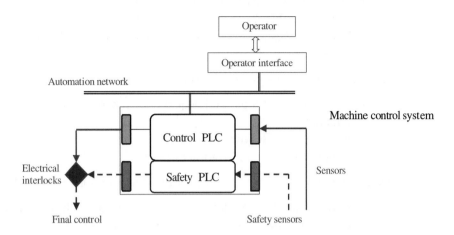

Figure 8.7
Machine with separate safety and control functions within a combined safety and control PLC

8.2 Benefits and disadvantages of safety PLCs

8.2.1 What are the potential benefits of using a PLC for safety?

Before we go any further into the design of safety PLCs, let's consider what benefits we are looking for.

We have just seen that there are some choices as to how the PLC can perform the safety-related functions. If we wanted to justify the PLC against any other method of implementing safety, what can it offer? Some of these are likely to be the same as the benefits of a standard PLC over hard-wired systems, such as:

- Software tools for configuration and management of the logic.
- Simplified wiring eliminates the problem of logic being embedded in the connections between hardware modules.
- Using software for safety functions allows the building of machine to be completed whilst the final protection logic is being developed. Facilitates late design developments and avoids wiring modifications.
- Standard control packages become cost-effective when machines are produced in quantities. Customized variations can be implemented with minimal costs.
- Centralized monitoring and display facilities.
- Improved coordination for large production lines or complex machine functions.
- Event recording and retrieval.

In particular, the safety PLC offers improved safety performance through:

- Improved management of safety functions. This is due to the strict control of application software through the programing tools. Unauthorized access to the software is prevented by password control. All changes required a double compilation task. All changes are recorded.
- Pre-approved software function blocks provide standardized methods for routine safety functions.

- Software function blocks are available to support setting up and tuning of specialized sensor applications such as light curtains. Muting applications can be developed more effectively.
- Powerful diagnostics for the detection of faults in the PLC and its I/O subsystems.
- Application blocks include for diagnostics to be performed on the sensors.
- Easier certification through the use of safety-certified PLCs with certified function blocks.

Productivity improvements will be sought for large installations through:

- More advanced logic and sequencing capabilities to reduce lost time in safety functions
- Better testing facilities
- Rapid detection and location of faults.

Bus technologies interfacing into some PLCs further improve cost-effectiveness through:

- Allowing plant-wide safety functions to be managed from central or remote stations.
- Remote I/O subsystems reduce cabling and reduce risk of cabling errors.
- Safety-certified field bus systems allow a single bus connection for all safety sensors on a machine, reducing wiring complexities.

8.2.2　What are the potential disadvantages of using a safety PLC?

Probably the main disadvantages are associated with capital cost when considering a PLC for a relatively simple application. Safety PLCs are much more expensive than standard PLCs and the cost of the software package and any special training must be added in. If the application is to be repeated many times the cost equation will become more favorable as the software investment is recovered. For a simple application the modular products we have seen will be a cheaper solution until the safety function becomes complex. For example, there are a number of packaged PLC solutions on the market for the complete safety functions for mechanical and hydraulic power presses. The scope of these solutions as a contribution to increased safety as well as higher productivity may well justify their capital cost.

Further reservations as far as the end user is concerned may arise from the need for more specialized knowledge on the part of the maintenance team. Again the benefits of improved diagnostics and testing facilities may offset this concern.

8.2.3　Development of safety PLCs

Why not use general purpose PLCs for safety functions?

Standard PLCs initially appear to be attractive for safety system duties for many reasons such as those listed here:
Attractions:

- Low cost
- Scalable product ranges
- Familiarity with products
- Ease of use
- Flexibility through programmable logic
- Good programing tools available
- Good communications.

But there are significant problems.

Problems:

- Not designed for safety applications
- Limited fail-safe characteristics
- High risk of covert failures (undetected dangerous failure modes) through lack of diagnostics
- Reliability of software (also stability of versions)
- Flexibility without security
- Unprotected communications
- Limited redundancy.

For machinery safety, in particular, the requirements for category 3 and category 4 demand a fault tolerance level of at least 1 (see Figure 8.8).

Single-channel PLC does not satisfy categories 3 and 4

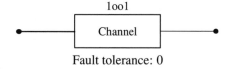

Fault tolerance: 0

Requires redundant channels to satisfy categories 3 and 4

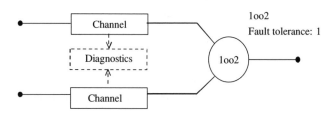

Figure 8.8
Fault tolerance levels showing 0 for a single-channel PLC and 1 for a dual-redundant system

The requirement for redundant channels requires a certain amount of specialized engineering and of course it doubles the cost of the PLC assembly. The further requirements in standards such as EN 954 for the safety system to self-test for faults before any restart (category 3) or for avoidance of an accumulation of faults (category 4) means that diagnostics must be provided.

So if we want to use a general purpose PLC it has to be specially engineered to meet the basic requirements for safety duties.

8.2.4 I/O stage diagnostics

Let's take a look at some basic problems for PLCs at the I/O stages.

Figure 8.9 is a simple example of the covert failure problem. The output stage of the PLC operates a fail-safe solenoid or motor trip relay. It may have to stay energized for weeks but we won't know if it is shorted until it has to trip the function. This is an unrevealed fail to danger condition or 'covert fault'.

The broken wire fault is an 'overt fault' or revealed fault, which will fail to a safe (off) state but creates a 'nuisance trip'.

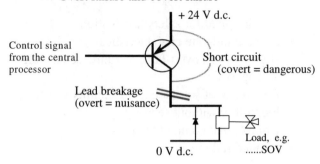

Figure 8.9
Failure modes of PLC output stages

Many users of standard PLCs have been able to introduce their own diagnostics for continuously self-checking the PLC during whilst it is on-line (e.g. see Figure 8.10).

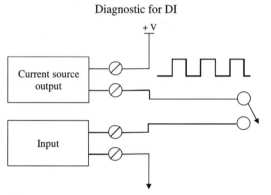

Figure 8.10
Simple diagnostic for PLC input stage

For output switching stages a typical method of self-testing consists of a pulsed off state that is too short to affect the load but which can be read back into an input stage as part of a test cycle (see Figure 8.11).

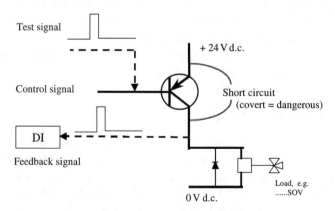

Figure 8.11
PLC output stage: cyclic test

An alternative approach for assuring input stage integrity is to use voting as a method of diagnosing a fault. In Figure 8.12 the digital input is not accepted as valid unless a majority of 2 out of 3 input channels agrees on the state (this is a 2oo3 architecture). If one channel disagrees with the other two the whole input stage can be treated as faulty or the fault can be reported and the PLC continues to operate on the majority vote until a repair is made.

Voting diagnostics for DI

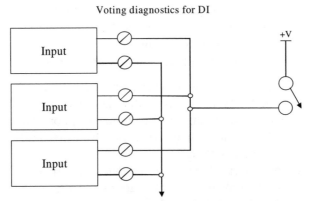

2oo3 voting is applied to diagnose a faulty input channel

Figure 8.12
2oo3 voting diagnostics for a digital input function

Overall PLC reliability

What we have seen so far are some simple measures to improve the safety integrity of the standard PLC. The problem for us is that the list of potential failure modes is a long one and the cost of avoiding or safely detecting each one is rising.

According to the process industry standard ANSI/ISA S84.01 there are at least 36 recognized types of failure modes in a typical PLC and another six failure modes in the I/O subsystems. Whilst some of these failures are rare, they are not acceptable for safety systems. What we are looking for is a situation where at least 99% of all possible failures can be certain to result in a safe condition for the machines and persons being protected by the PLCs. So on the grounds of hardware performance alone engineers are faced with a tough task to make the standard PLC do the job (see Figure 8.13).

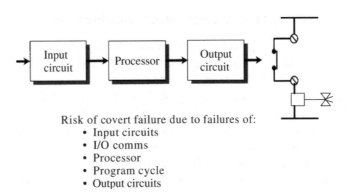

Risk of covert failure due to failures of:
- Input circuits
- I/O comms
- Processor
- Program cycle
- Output circuits

Figure 8.13
Basic PLC architecture without diagnostics type – 1oo1

8.2.5   Software reliability considerations

There are similar reservations about the suitability of standard PLC software for safety:

- Potential for systematic faults
- Program flow and monitoring is not assured
- Too much scope for random applications (high variability)
- Lack of security against program changes
- Uncertain response in the presence of hardware faults
- We can't test for all foreseeable combinations of logic
- The failure modes are unpredictable
- Re-use of old software in new applications.

Fundamentally the problem with using software in safety system lies with the potential for systematic faults, i.e. faults that are not random such as component burnouts but have been introduced during the specification, design or development phases by errors in those processes. Such faults may then lie dormant until just the right combination of circumstances comes along.

One of the aspects of software that makes it particularly difficult to control is the ability for it to be re-used in new applications not originally intended by the designers. The tendency to use 'cut and paste' techniques to make up a new program creates a risk that the wrong features have been introduced to a new product.

It would help if it were possible to detect all systematic errors at the testing stages but it is well understood that there will be many combinations of logic and timing that cannot be fully explored in testing without a prohibitive cost in time and labor.

8.2.6   Do all these objections eliminate standard PLCs?

No they are not eliminated, but they are likely to be the most expensive route to go. There are many existing applications where standard PLCs have been adapted to safety-related control functions. These applications have involved installing dual-redundant sets of PLCs and a number of self-checking measures along the lines we have shown. In some cases the recognized certification bodies for a specific application have approved these solutions.

However with the availability of specially engineered safety PLCs for general use in industry it becomes far less attractive on grounds of cost and engineering effort to take that route. Even when the system has been adapted for safety it still requires a third party to verify that it is suitable for the safety duties.

8.2.7   Upgrading of PLCs for safety applications

Summarizing the position   It is possible to upgrade software engineering through improved QA techniques. It is possible to consider dual-redundant standard PLCs in hot standby mode but the standard PLC does not lend itself to covering all the possible failure modes through the normal fault detection systems. We can add our own diagnostic devices for some types of failures as we have seen. But at the end of all these extra efforts we have the problem that we have built a special application that needs to be carefully documented and maintained. And then we have the problem of proving it to others or certifying it for safety duties.

In a review of the position regarding standard PLCs compared with safety PLCs, industry specialist Dr William M. Goble (from a paper by Dr William M. Goble, Exida, www.exida.com) concluded:

> *The realization of many users that conventional controllers cannot be depended upon in critical protection applications creates the need for safety PLCs.*

The standards are high for safety PLC design, manufacture and installation. Anything less than these high standards will soon be considered irresponsible, if not negligent, from a business, professional and social point of view.

This is basically the case for vendors to produce a special purpose PLC built specifically for critical safety applications. Let's look at what it takes.

8.3   Characteristics of safety PLCs

Now let's summarize what we can expect from a safety PLC and take a look at some of the ways that vendors have met the challenge.

In summary   The answer to the problem of undetected faults in PLCs lies in the concept of fault coverage and fault tolerant systems. The answer to the problem of hidden defects in software is high-quality embedded (i.e. operating system) software combined with strictly defined and constrained user-programing facilities.

8.3.1   Hardware characteristics of a safety PLC

- Automatic diagnostics continuously check the PLC system functions at short intervals within the fault-tolerant time of the process.
- High diagnostic coverage means that at least 99% of all hardware faults will be detected and notified for attention and repair.
- Provides a predictable and safe response to all failures of hardware, power supplies and system software.
- Redundant hardware options available to provide safe operation even if one channel has failed.
- Fault injection testing of the complete design is performed to ensure safe failure response to all known faults.
- I/O subsystems continuously check all signal channels. I/O bus communications are self-checking; faults result in safe isolation of affected I/O groups.
- High security on any reading and writing via a digital communications port.

8.3.2   Software characteristics of a safety PLC

Software quality assurance methods are deployed throughout the development and testing of both operating system and application software development. Software development takes place under 'safety life cycle' procedures as specified in IEC 61508, part 3. The software design and testing is fully documented so that third-party inspectors can understand PLC operation.

- Operating system uses a number of special techniques to ensure software reliability. These include:
    - 'Program flow control' checking, this insures that essential functions execute in the correct sequence
    - 'Data verification' stores all critical data redundantly in memory and checks validity before use.

- Operating system and user application software tools are approved for safety by third-party approval bodies.

- Operating system and programing package supplied by same vendor as the hardware.
- Software and hardware integration tested by approval bodies.
- Extensive analysis and testing carefully examines operating systems for task interaction.
- Application software uses 'limited variability languages' to restrict to end users to working within a framework of well-proven instructions and function blocks.
- All application software updated transparently to redundant channels.

Whilst all of the above are general performance and qualification features of the safety PLCs, the practical end user will also be interested in some more down to earth characteristics. For example, users will look for:

- Economically priced PLCs at the right size for typical machine applications
- Input channels suitable for all common safety sensors and output channels suitable for connection to secondary or final control contactors or solenoids
- Remote I/O capabilities to allow input and output modules to be mounted close to the parts of the machine or production line that they serve
- Speed of response fast enough to deliver E-stop and safety trip responses without increasing risks to persons
- Low software engineering costs, library of certified safety function blocks
- Easy to program with fill-in-the blanks function blocks plus simple ladder logic or sequential logic instructions. Program language should be as close as possible to the type in use for basic control PLCs
- Good testing and diagnostic facilities
- Rapid identification of faulty parts and easy replacement
- Compatible but safe connections to automation control networks.

In other words it will be best if the safety PLC is, in all respects except safety, the same product for the end user as the standard PLC.

8.3.3   Design of safety PLCs

This section illustrates some of the features typically found in safety PLCs. We begin with a single-channel system and work toward more sophisticated designs. We can only provide here a brief look at the key features and would advise that it is generally possible to track these features in greater detail by close study of the technical descriptions of the products provided by the various manufacturers.

We shall see in this section that different types of safety PLCs are evolving to suit the type of industry applications. The major division is between process industry applications and machinery safety.

8.3.4   Single-channel safety PLC architecture with diagnostics

Figure 8.14 shows the basic architecture of the PLC upgraded to include for diagnostic devices embedded in the construction of the PLC. This unit is able to overcome the objections listed for standard PLCs and is now the basic module concept for several safety control system manufacturers. Essentially this unit in single-channel configuration will trip the machinery if any of its diagnostic functions finds a fault in any of the stages: input, CPU, power supply or output.

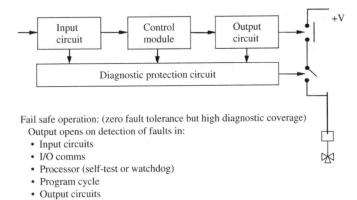

Fail safe operation: (zero fault tolerance but high diagnostic coverage)
Output opens on detection of faults in:
- Input circuits
- I/O comms
- Processor (self-test or watchdog)
- Program cycle
- Output circuits

Figure 8.14
Single-channel safety PLC architecture with diagnostics type – 1oo1D

The term 1oo1 comes from the notation that any one dangerous fault in the single-channel system will cause a failure of the safety function. The D denotes diagnostics protection.

There are two important features to note here:

1. Diagnostics in the single-channel system must be performed within the PLC cycle time such that the 'fault-tolerant time' or 'machine safety time' is not exceeded.
2. The diagnostic circuits must have a means of shutting down the outputs of the PLC that is independent of the output switching circuits. This is sometimes known as a 'secondary means of de-energizing' (SMOD).

8.3.5   Fault-tolerant time

The principle of fault-tolerant time or machine safety time is illustrated in Figure 8.15.

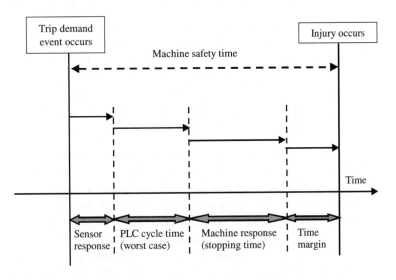

Figure 8.15
The machine safety time

In Figure 8.15 the PLC cycle time must be short enough to ensure that the safety response is always fast enough to stop the machine before the injury event can occur. If the PLC is to have a part of its cycle taken up with self-testing routines (diagnostics) these must not extend the cycle time such that there is no safety margin in the time frame before the next cycle.

This is where there is a point of difference between the process industries and the machinery world.

- Fault-tolerant times in the process industry are typically greater than 1 s
- Fault-tolerant times for the machines in industry are typically less than 1 s.

Since the stopping time of a machine takes up much of the fault-tolerant period the safety PLC is required to have a fast response. Hence single-channel architectures are not suitable for machine safety PLCs due to the limited time available for comprehensive diagnostics.

In process applications or others where the fault-tolerant times are well above 1 s it is quite possible to achieve high coverage levels in the allowable time frames. Hence single-channel PLCs are available for lower integrity applications (typically up to SIL-2).

8.3.6 Diagnostic coverage

An essential requirement of the diagnostic tests for the PLC is that they should cover a high percentage of potential faults. Diagnostics linked to appropriate control actions convert potentially dangerous hidden faults into safe-mode failures of the PLC. If there remains a substantial number of a potential fault that cannot be detected by diagnostics then much of the benefit is lost.

Diagnostic coverage is the ratio of safely controlled faults to all possible faults

From Steven E Smith: Fault coverage in Plant protection Systems, ISA 1990 Paper #90

This paper provides an excellent review of fault coverage techniques.

8.3.7 Dual-channel safety PLCs

There are strong incentives to design dual-channel 1oo2 architectures for safety PLCs:

- In machines the concepts of categories 3 and 4 require at least fault tolerance level 1, hence redundancy is essential.
- If diagnostics require more than the fault-tolerant time of the functions, then a redundant channel will provide protection until the faulty channel is detected.
- In machines or plants where a crash shutdown will cause losses, the redundant channel allows the plant to continue operating for a limited period whilst the faulty channel is repaired, thus avoiding a 'nuisance trip' and lost production.

In this version the entire logic solver stage from input to output is duplicated and if one unit fails its diagnostic contact will open the output channel and remove that unit from service (see Figure 8.16). The SIS function then continues to be performed by the remaining channel.

The notation 1oo2D applies because the system will still perform in the presence of one fault amongst two units. The parallel connection of the two units substantially improves the availability. Note that diagnostic performance is further improved by cross-linking between the CPU of one channel and the diagnostics of the second channel. The voting performance achieves 1oo2 because if either channel detects a trip command the other channel is forced to follow. This arrangement is popular with process industries due to its avoidance of nuisance trips.

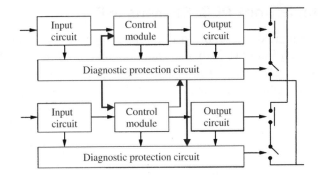

Both channels must operate to trip output. Reverts to 1oo1D if
module fault is detected. Diagnostics must check other CPU

Figure 8.16
Dual-redundant channel safety PLC architecture with diagnostics parallel-connected type: 1oo2D

However for machinery applications we are less interested in the avoidance of nuisance
trips and it is likely that the series connection of the two PLC output channels will satisfy
category 3 and 4 requirements.

8.3.8   Series-connected PLCs

The two units are operated in series as shown in Figure 8.17. This configuration is
preferred for machinery applications since we are not concerned so much with losses due
to 'nuisance trips'. Series connection of the outputs means that either of the channels can
trip the plant.

There are a number of possible variations on the 1oo2 architecture PLCs. For example,
the I/O subsystems can be constructed as 1oo2 units capable of being commanded from
either of the CPU or control module stages.

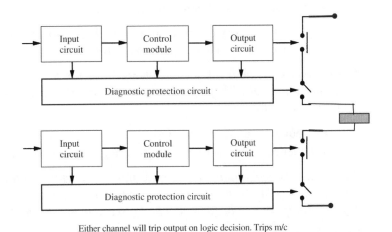

Either channel will trip output on logic decision. Trips m/c
if either module is faulty. High-integrity version

Figure 8.17
Dual-redundant channel safety PLC architecture with diagnostics series-connected type – 1oo2D

8.3.9 Example: 1oo2 PLC: Siemens 95F

One example of the 1oo2 PLC for machinery safety is the Siemens S5-95F. This unit is constructed from two identical subunits of the S5-95U PLC. Whilst it has 1oo2 architecture it is designed to operate as a single unit for machine safety applications. Hence it fits our model for option 1 as shown in Figure 8.18.

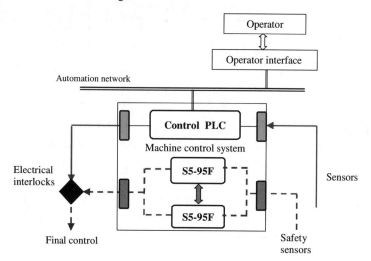

Figure 8.18
Machine with 1oo2D safety PLC

The figure shows the S5-95F installed as safety-related control system with a separate PLC for basic controls. It is possible to run some basic controls in the S5-95F and if the capacity of the unit is sufficient for all controls the basic unit will not be needed.

Each unit comprises a processor module and a small number of I/O channels as follows:

- Sixteen safety-related digital inputs
- Four safety-related hardware interrupts
- Eight safety-related digital outputs
- Eight non-safety-related digital outputs (4 per subunit)
- Two safety-related hardware counters.

Some expansion capacity can be achieved by linking to additional I/O modules. Siemens describes the mode of operation as follows:

During operation, both sub-units exchange data for comparison, quickly and reliably via fiber optic cable. They operate in synchronism with the same user program and cyclically compare:

- Input and output signals and
- Other relevant data, e.g. results of logic operations, flags and counters.

The control system automatically goes into a safe stop condition for different and/or internal errors/faults. Additional error responses, e.g. de-activation (passivation) of defective outputs, can be defined when programming the control system.

The 95F is presented as a small control unit that always operates as a dual-channel unit. Hence it does not offer the option to keep running on one channel if a fault occurs in the other. For 'high availability' applications a second 95F pair would be used with parallel connection arrangements similar to our 1oo2D diagram above.

Programing of the S5-95F is done with a safety-certified subset of the well-known S5 programing language and hence many users will already have the experience of the language from general PLC applications. In addition

Libraries with tested program blocks, which have been accepted by the relevant testing organizations, are available for many applications. They only have to be downloaded into the S5-95F, called up and parameterized. This significantly reduces engineering costs. Further, user-friendly parameterizing, configuring and diagnostic functions of the standard STEP 5 programming package significantly simplify commissioning.

This unit has been available since 1993 and has been widely used in machinery applications.

8.3.10 Other 1oo2 examples

In the small modular range of 1oo2, Rockwell has the AB GuardPLC units in two sizes:

1. *GuardPLC 1200:* Small fixed size unit with digital inputs and outputs. It has 20 DI, eight DO, two counters, self-testing and with multiple I/O test points but without analog capability. Said to be suitable for safety applications up to SIL-3. The architecture of the unit is unclear at the time of writing but we assume it has a dual-redundant processor system if it has a SIL-3 rating.
2. *GuardPLC 200:* Larger framed unit with optional I/O modules. Digital and analog capability. Rated for SIL-3 applications. Architecture is also unclear.

Both the above units are programed with a safety-certified subset of the RS Logix programing language. As with the Siemens unit this means that users will already be familiar with most aspects of the programing.

8.3.11 Single-channel safety PLCs

More recently manufacturers have been introducing single-channel non-redundant safety-certified PLCs to serve the smaller and simpler machinery applications found in large numbers in factories. The IEC 615108 standard sets out the requirements for safety integrity in PES hardware in a very flexible way. For SIL-1 and SIL-2 applications it is acceptable to operate with single channel architectures provided the level of diagnostic coverage is sufficiently high. This has encouraged the growth of small single-channel safety controllers.

8.3.12 Larger 1oo2D systems

In the process industries 1oo2D systems are popular for their ability to meet high-integrity ratings whilst offering high availability for production. Leading products in this sector include:

- Honeywell FSC systems
- Hima Hi-Quad
- Siemens-Moore Quadlog.

These controllers have high capacities in terms of I/O count and processing power. These units also offer options for dual-redundant I/O subsystems or for single I/O subsystems dependent on the integrity levels being sought. However we have already noted that process industry systems are not designed for the fast responses required by many machinery applications.

We will not detail these systems here, as they are not usually used for machinery applications. However it is worth noting a significant design feature that has been introduced in some of these systems. These manufacturers now offer what is sometimes called Quadruple Redundancy. This is 2oo4D architecture as shown in Figure 8.19.

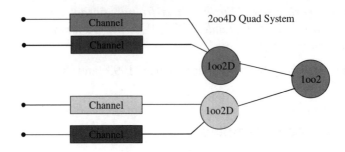

On detection of one faulty channel the failed pair is removed and the remaining system still has fault tolerance 1.
Still qualifies for SIL3 during repair

Figure 8.19
Architectures for safety systems – 2oo4D

In these designs the PLC units are made in dual-redundant pairs on a single frame or module. When one dual unit fails, the second unit can remain in service indefinitely and still provide SIL-3 integrity and satisfy category 4 because it is still a dual-redundant unit.

For machinery and automation applications there is the recently introduced Siemens 400F system. This PLC system is based on the dual-redundant H series processors and is offered in a single-channel or dual-redundant PLC format. Figure 8.20 is based on material published by Siemens.

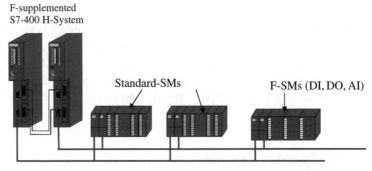

Standard PROFIBUS DP peripheral connection with fail-safe
DP-communication (ProfiSafe)

Figure 8.20
Siemens S7-400F safety PLC

For a detailed explanation of this system we would refer the workshop attendees to the Siemens Application Manual, Chapter 4. The important features to note here are:

- The processors are designed to allow non-safety-related functions to be processed as well as safety-related functions. Functional separation between the two duties is assured.

- This system is designed to use a fail-safe version of the PROFIBUS DP field-bus communication system to provide access to remote I/O subsystems (known as SMs).
- The SMs are available in non-safety (SM) and in safety versions (F-SM).
- The F-SMs have their own internal system diagnostics similar to those found in a centralized safety PLC. They have the ability to shutdown if a fault is detected.
- The F-SMs will also shutdown if the 'ProfiSafe' communications fail. Intensive self-testing and timeout protection in the communications processors ensures that any malfunction in the bus signaling will cause a shutdown of the F-SM.
- The intention is to allow ProfiSafe communications and standard Profibus to co-exits on the same network. Hence a plant-wide Profibus network can serve both basic control and safety control duties.

This system is intended for large automation applications such as those found in the motor industry or for process control applications. More recently Siemens have introduced a smaller version of this concept designated Simatic S7-300F and using fail safe signal modules designated ET 200S. These developments indicate a significant trend towards integrating safety and control system functions into a single control system package.

8.3.13 Common cause potentials in redundant PLCs

Despite the practice of high-quality hardware and software engineering there is always a small possibility that two identical PLCs connected in redundant mode will both suffer the same common defect. Whilst this is possible for identical hardware modules there is also the possibility of a systematic fault in the embedded software.

Manufacturers with identical processor systems have resorted to various design measures to try to minimize common mode errors and it appears that they are able to achieve satisfactory results (e.g. the use of switching between calculate and verify mode in the Siemens-Moore Quadlog PLCs). However, it remains as a good design principle in safety systems to introduce diversity of hardware and software into redundant systems.

This concern was taken into account by PILZ in the development of their PSS safety PLC system that is specifically designed for machinery applications. We shall look at the basics of the PILZ system in the next section.

8.3.14 Triple modular redundant (TMR) systems

The safety systems built on the 1oo2D modules have found a strong market in the general area of process plant applications. In some of the most demanding safety areas including offshore oil and gas, and in the nuclear field these systems have to compete with the alternative architecture based on the principle of two out of three voting. These are known as triple modular redundant (TMR) systems. Figure 8.21 illustrates the principle.

These units have the advantage of not having to use such complicated diagnostics as the 1oo2D systems because every stage of the PLC can be built up with two out of three majority voting for each function.

For example, the three input stages operate in parallel to decide which is the correct value to pass to the processor. The three processors compare data and decide which is the correct action to pass on to the output stages, etc. All internal communications can use a triplicate bus.

A good example of this type is found in the range of Triconex Safety Controller products. Triconex systems are widely used in the power machinery field for the protection of large compressor sets where high speed of response is needed as well as high safety integrity and high availability.

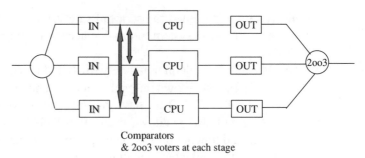

Comparators
& 2oo3 voters at each stage

No single point of failure. High safety integrity. High availability

Figure 8.21
Triple modular redundant (TMR) PLC

Key features of TMR safety systems:

- No single point of failure
- Very high safety integrity
- High availability
- High speed of response
- Transparent triplication
- Delayed maintenance capability (for remote or unmanned locations)
- Hot spare capability for I/O modules
- Sequence of event recording.

TMR systems are generally found to be expensive compared to 1oo2 systems due to the need to build special purpose CPUs and I/O subsystems. Those products that can use slightly modified versions of standard PLC hardware may show cost savings over TMR designs.

8.3.15   PILZ PSS safety PLC: 1oo3 architecture

The PILZ safety PLC has some things in common with the TMR system but there are also major differences:

- It seeks to avoid common mode problems by using three different models of CPUs made by three different manufacturers.
- It also has a different objective from the process industry TMR designs because it concentrates on making a fail-safe response by only allowing the safe outputs to remain on if all three CPUs are healthy and only if they agree the required output states of the system.
- It does not provide for the machine to carry on running when processor has failed.

These last points underscore the difference in objectives between 2oo3 and 1oo3 voting systems.

The PILZ PSS system can be regarded as a 1oo3 architecture system with non-identical redundant CPUs. There may be some confusion here over the notation because the PILZ handbooks describe the unit as having a '3 out of 3 (3oo3) voting system'. This terminology appears to arise from the German notation that calls for 'number of channels that must agree for the output to stay on'. In this case three channels must all agree. The IEC notation we have been using calls for the 'number of channels that must call for a trip for the output to trip'. This results in the term 1oo3 for this PLC.

8.3.16   Key features of the PILZ PSS

The PILZ PSS systems have had a major impact on the machinery sector usage of PLCs for many years since their introduction in the mid-1990s.

For a detailed description of this safety system we recommend the PILZ Publication: 'Guide to Programmable Safety Systems', Volume 2. Our notes here are based on information in the PILZ manual.

The basis of the PSS controllers is the three-channel voting structure we have described above. Figure 8.22 shows the arrangement.

Structure of the PILZ PSS safety systems

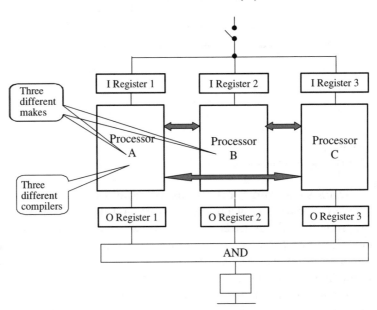

Figure 8.22
Structure of the PILZ PSS controller

The system comprises three separate controllers, each one different from the other and supplied from different makers. Each controller has a different operating system and has its own code compiler written by different companies. These measures reduce (practically eliminate) the chances that a systematic error or hardware fault will occur as a common fault in all three controllers. The PILZ description then continues:

> *Data is processed in parallel via the three controllers. Comparison checks are performed between each controller. Each processor has its own input and output register. The output register in each device is compared in an AND gate. An output will only be enabled when all three agree. This is the case for all "bit" and "word" functions. In other words, the PSS acts as a 3 out of 3 (3oo3) voting system.*

The PSS system has been designed to allow standard (or basic) control functions to be performed as well as safety-related control functions. Safety functions are known as FS and standard are called ST. FS software runs in all three processors using an FS working bus. A separate bus is provided for the A processor to perform ST instructions.

Figure 8.23 shows a diagram of the PSS system structure with processor A performing the safety and standard functions.

The dual-function structure of the PSS has fail-safe controls operating with an independent triple voting bus whilst standard or non-safety functions operate only in processor A on a separate working bus.

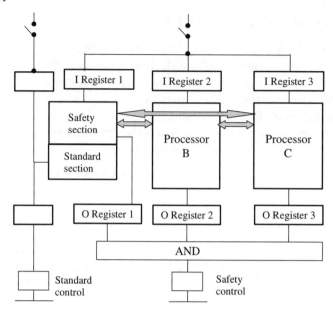

Figure 8.23
PILZ PSS: Standard and safety control

The FS safety programs are always able to operate independent of the ST section. The ST section is programed as normal PLC. The FS section applications are programed with certified safety software function blocks and are automatically compiled into three versions by the programing system. Hence this triplication of the application code is transparent to the end user, which is a normal case for all the redundant safety PLCs.

We shall see in the next section that processor A is also the managing processor for the PSS systems because it is in charge of the synchronizing of each scan of the processors.

The PILZ PSS safety PLCs can be seen to satisfy the options shown at the start of this module for a combined standard and safety PLC unit as in our option C. It can of course also meet option B.

The method of self-testing employed in the PSS is of interest because it differs from other safety PLCs by running its cyclical tests in thin slices spread over many program cycles. Here is a summary of the testing sequence based on the PILZ description:

- On power up, the PSS runs an internal check of all its functions, testing the internal system software, firmware and hardware. The check sums of the three memories are compared together with the status of the external and internal power supplies. This is also where the dual pole outputs are tested for operation. This pre-start check takes about 40 s.
- Peripheral devices are then checked.
- If all tests are passed the FS section is allowed to run.

Cyclical program checks begin. For each cycle the following are monitored and deviations from normal result in shutdown,

- Program block run time
- I/O status checks
- Power supplies
- Test slices the 40 s pre-start check.

Each program cycle performs a slice of the pre-start checks until all checks have been completed. PILZ states that the pre-start checks employ about 40 000 test slices of about 1 ms each. This means that the testing can be executed in small steps each time the program scan is completed. The result is that the entire system check can be completed over a period of typically several hundred seconds (see Figure 8.24).

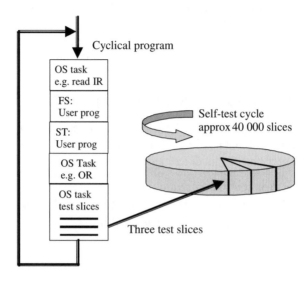

Figure 8.24
Test slicing in the PSS

The testing regime is adjustable. So, for example, if 10 test slices are executed per cycle this will add 10 ms to the cycle time. Typically a 50 ms scan interval is thereby increased to 60 ms. This will require 4000 program cycles to complete the 40 000 test operations and the task will be completed in 240 s.

The time slicing of the tests means that the PILZ PSS systems can achieve a fast program scan rates and provide a fast response to any safety trip condition. The possibility of fault being undetected for a period of say 6 min up to perhaps 30 min is acceptable because the fault is only likely to be present in one of the three processor units. The remaining two units will still provide adequate safety integrity until the fault is detected. The possibility of the fault being a common mode across the three processors is extremely low due to the diversity of hardware and software employed in the design.

8.3.17 Conclusions on processors

Without going any further into the details of safety PLCs it is clear that there are a fair number of devices available to the project engineer wishing to use them.

Here's a summary of some of the features we have seen so far:

- Intensive self-testing ensures that a very high percentage of possible dangerous faults are detected.
- In machinery systems, detection of a dangerous fault usually results in immediate shutdown of output signals to a safe condition.
- In process automation the preferred reaction to a fault is to keep running with reduced integrity and make repairs quickly.
- The shutdown is assisted by using a secondary means of de-activation to overcome any short-circuited output stages.

- Redundant processor systems are standard practice for the higher integrity systems.
- Identical redundancy is commonly used but is supported by intensive self-testing.
- Diverse redundancy reduces the risk of common mode faults.
- Triple redundancy with diversity reduces the diagnostic workload and speeds up scan rates but complicates manufacture.
- Small modular safety PLCs are available for single machines.
- Larger safety PLCs are available for automated machinery systems.

8.3.18   What else does the PLC have to provide?

Now that we have established some basic knowledge of safety PLCs, we need to look at additional attributes needed to put them to work in a practical machinery safety system. In the next sections of this chapter we will consider:

- Application software
- Safety-certified bus systems for plant signals
- Communications and networking
- Certification.

8.4   Application software

Once the PLC system has been chosen and the hardware has been specified to suit the application what really matters to the end user is:

How do I get the programs up and running?

We have already met the concept of 'limited variability languages', which in practice means high-level application languages (e.g. statement list, ladder logic) and pre-written function blocks. The function block approach seeks to anticipate most things that we would want to do in machine safety techniques and provides pre-written and thoroughly tested modules of software called function blocks. The user is left to assign names, channel addresses, times, etc. and to customize the standard actions of the selected function block.

This approach is made easier by the existing functional safety requirements for particular types of machines that we know, for example, as type C standards. The same applies to familiar safety devices such as two-hand controls, E-stops and light curtains with muting or single-break/double-break initiation. These function blocks are fully integrated with the safety PLC so that all test pulses, dual-redundant signal checks, earth and short circuit checks and all restart sequences are handled within the block functions. This of course, makes the function block equivalent to the hardware monitoring devices we saw earlier in the workshop.

It appears that PILZ, in particular, have developed several function blocks dedicated to standard machine safety procedures. Their range includes:

- SB 061 E-stop monitoring
- SB 067 feedback loop monitoring
- SB 068 ESPE/AOPD operations for start and stop of a machine
- SB 080 drive and monitor for a press safety valve
- SB 087 monitoring the function of a single hydraulic valve.

Each of these blocks will be reviewed and tested by an approved certification authority, e.g. in machinery applications in Germany: BG.

The safety control program can then be constructed as say a statement list from which the various blocks can be called. The blocks are configured for their tasks by assigning the correct I/O channels and input variables.

For example, the safety gate-monitoring block will be used to generate secure 'permission to start' signal that will be linked to a next block that will lead to a motor starting or hydraulic valve being opened.

Programs are constructed on a graphical interface on a programing PC workstation. The workstation typically permits simulated testing of the operation before the correct version is submitted for compilation.

8.5   Safe networking

The next step in the growth of machinery safety systems has been the introduction of safe networking. This evolution follows the same rationale as for any automation network, the attraction being that it increases the scope and power of any control system without adding the complexity of multiple wiring connections in hardware.

The development of safe networks has been based on the principles laid down in the programmable system standards DIN 19250 and IEC 61508. These standards focus on achieving high integrity through intensive diagnostics and high-quality software engineering combined with fail-safe techniques. Networks are able to comply with the performance requirements defined by these standards for SIL ratings even in single-channel designs.

Clearly there are some essential requirements for a field-bus network to be used in safety:

- The network must guarantee delivery of data packets in a minimum time frame.
- All devices connected to the network must be guaranteed to fail to a safe condition if a network fault is detected.
- Network diagnostics must be capable of detecting virtually all (more than 99%) conceivable errors in sufficient time to avoid a dangerous fault being present on the network.
- All devices on the network must be known to be present and must be polled to confirm their safe status.
- Single-channel networks must deliver good availability to avoid spurious downtime.
- Dual-channel networks should be available as options for high availability systems.

Here are some examples.

8.5.1   The PILZ safetyBUS-p

SafetyBUS-p allows up to 32 subsystems to be linked and interfaced to a safety PLC. Local safety interlock functions are performed in the subsystems using safety-certified program blocks (see Figure 8.25). Peer to peer communications permit safety functions to operate without a host.

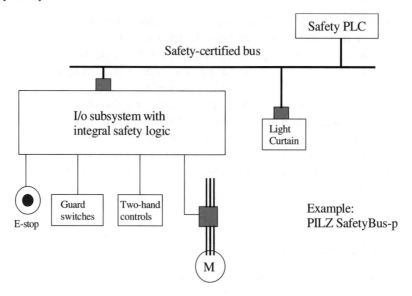

Figure 8.25
Simplified representation of PILZ SafetyBUS-p

The following is an extract from the manufacturer's description of the product:

SafetyBUS-p is based on an event-driven bus procedure, i.e. data is only sent when the status at the I/O or field module has changed. This means that SafetyBUS-p is particularly suitable for networking plants that combine functions with variable signal frequencies and fast reaction times.

SafetyBUS-p is a multi-master system based on the proven CAN bus system. 64 subscribers can be connected via SafetyBUS-p using the PSS-range of programmable safety systems. Subscribers may include not only the PSS programmable safety systems but also decentralized I/O modules or field modules (e.g. light curtains) that are connected to SafetyBUS-p directly. Cable runs of up to 1,000 m can be installed.

The strong features of this system have the ability to place input/output modules (eight inputs/eight outputs) at convenient locations in an automation plant. The standard safety devices we have seen such as E-stops, two-hand controls and light curtains can then be connected at those points.

The remote I/O stations allow the software function blocks used in the standard safety controller (PSS) to be used with the distributed devices. Bus technology also allows for remote devices such as light curtain controllers to be configured from the central controller for automation applications such as muting and blanking of beams.

One of the major advantages of this type of product for safety in automation is the ability to perform diagnostics on the sensor subsystems and hence manage any fault conditions efficiently and with minimum loss to production.

8.5.2   AS interface bus

The 'actuator-sensor' bus system has been established for several years in basic machinery control and the concept has been successfully extended to allow safety-related functions to share the same network without risk of compromising the safety functions. It has been certified by testing authorities to be capable of providing safety functions up to category 4.

The AS interface allows single-cable networking of safety and non-safety devices with data transfer to plant level controls. Safety monitors are parameterized to carry out safety functions and act as hosts for a cluster of devices. Multiple networks can be linked for multistage operations.

As we noted before the ability of networks to handle safety-related data is recognized as valid within the requirements of IEC 61508 because they are able to achieve a very high level of diagnosis for faults. This is technically known as high diagnostic coverage. Provided the design of the network and its hosts can achieve a safe response to any detected failure of the sensor, its interface to the network or in network itself can be considered to be as safe as hard-wired device.

Figure 8.26 is a simplified one intended to show that an E-stop and guard switches could be logically linked to trip a bus connected drive control.

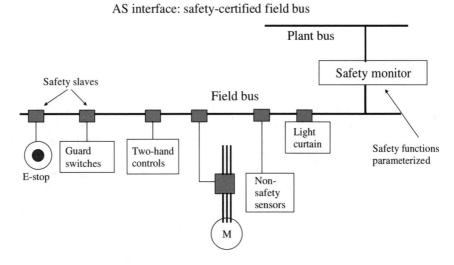

Figure 8.26
Simplified representation of AS interface bus handling E-stops and other safety functions

Here is an extract from the manufacturer's description of the technical aspects of the AS interface:

With the Safety at Work system expansion, AS-Interface is again setting new standards. More than just a safety-related bus is created, as safety-related and standard data are transferred together along a yellow AS-Interface cable. A response time of max. 35 ms ("worst case") sets new standards when it comes to safety-related field buses.

The components for Safety at Work are, in compliance with EN 50295 and IEC 62026-2, fully compatible to all of the other AS-Interface components. Transfer protocol forms the basis for secure data transfer. Unique code table is saved in each slave, which allows the master to identify them. When a safety-related network is powered up for the first time, the safety monitor recognizes the safety slaves and saves the associated slave-specific code tables in a comparator. Each time the master calls a slave, the comparator checks the actually sent code values against those that it has stored to ensure that they correspond. If deviations occur or monitoring times are violated (watchdog), safe shutdown is initiated at the safety monitor via dual-channel enable circuits. Information can be additionally evaluated for diagnostic purposes using the plant or system control.

8.5.3   Profisafe/Profibus

We saw in Section 8.3.9 that Siemens offer the 400F safety system with a shared Profibus and Profisafe network linking to I/O subsystems. In this design fail-safe and basic control system components share the same network. Some of the features of this plant-wide bus design include:

- Safety devices can connect directly to the network with fail-safe communications ports.
- All safety devices fail-safe on loss of communications.
- Safety function messages are event-driven and hence have rapid response.
- Fail-safe I/O subsystems ensure safe shutdown on loss of communications.
- Dual-network version are offered for process control to ensure high availability against spurious shutdowns.
- Safety Integrity Level 3 (IEC 61508), Category 4 (EN 954-1) and AK6 (DIN V 19250) have been achieved.

8.5.4   General practices for safety networks

When field bus systems are used for safety applications it is likely that they will be applied to complex machines or an array of machines assembled for a manufacturing or assembly line. As with safety PLCs the designer's task will be to identify individual safety functions within the overall scheme and arrange a hierarchy of control so that the system can be managed effectively.

From the point of view of safety, it may be sensible to partition the plant into subsections. This takes into account the ability to form groups within the safe bus system enabling safety-related data from a whole plant to be controlled through a single safety bus. Subsections, however, may be assigned to different groups. Should a fault occur, only the respective group would need to switch to a safe condition. It is also possible to form supervisory groups.

A typical example of this would be in E-stop applications, where the E-stop function is valid for the whole plant, irrespective of the location of the relevant E-stop button. In a case such as this, where a signal group has been formed, the E-stop function must be operative for the whole plant, even if a section of the plant (signal group) is in a group stop, e.g. for maintenance work.

8.6   Classification and certification of safety PLCs

Having looked at some safety PLCs it should be clear that the technology is well established to meet the objectives of achieving high-safety integrity using programmable electronic systems. The remaining obstacle is the task of proving the performance of the product. This is where the earlier DIN 19250 standard and the new IEC 61508 standard make a major contribution by setting out engineering procedures and technical performance standards that must be met in order to achieve designated SIL performance levels.

We have noted that certification must cover the hardware and diagnostic performance capabilities as specified by IEC 61508. Certification must also cover the software engineering life cycle activities leading to the embedded software as well as the programing tools supplied for the end user.

The combination of hardware and software certification means that the end user can obtain a fully certified product for the logic solver section of the SIS. This does not of course relieve the end user of any obligations as far as the quality of his software application/configuration is concerned.

Certification is a comprehensive and specialized task undertaken only by well-established authorities such as TUV. Details of testing guidelines and practices are outside of the scope of this workshop but it is helpful to note that TUV publishes details of their testing programs on their website. In particular their website publishes a list of type approved systems so that the progress of certification for each manufacturer's product can easily be checked.

8.7   Summary

We have seen that there are strong objections to placing safety-related control systems into standard PLCs.

The difficulty faced in many industries at present is that some plant engineers are adding safety-interlocking functions into existing PLC-based control systems by connecting safety devices directly into the standard PLC. This is convenient, cheap and appears to be a logical thing to do. It is in fact potentially dangerous because:

(a) The separation of safety and control functions is lost. There is the risk that failure or corruption of the standard control system software or the hardware will result in a hazardous condition and at the same time the safety function will fail.

(b) The integrity of the safety function is compromised by use of a logic solver with unprotected software, an absence of self-testing and an unpredictable response to hardware faults.

Technology options for the logic solver have been examined and it has been established that the safety PLC has succeeded in overcoming the inherent drawbacks of basic PLCs when applied to safety systems.

All the features that have made PLCs attractive to machinery control engineers are becoming available in safety-certified forms. This enables the development of more efficient control schemes incorporating safety-related functions within the automation scheme. For medium and large automation systems the integration of safety systems into the overall control system is now feasible without compromising the essential independence and integrity of the safety functions.

References

[1]   Siemens Safety Integrated Application Manual, Chapter 2.
[2]   PILZ: Guide to Machinery Safety.
[3]   PILZ Guide to Programmable Safety Systems, Vol. 2, IDC Safety Instrumentation Manual.

9

Introduction to standards for programmable systems

9.1 Introduction

The growth and widespread application of programmable systems such as PLCs and microprocessor-based sensing systems requires an appropriate set of engineering codes and practices similar to those that have served the hard-wired solutions of earlier days.

Until recently design engineers have had to find ways of interpreting the earlier hardware-based safety design standards when trying to apply new programmable equipment in safety systems. The older standards made very little provision for the new technologies leaving engineers with a difficult choice. Either:

- Do not use programmable equipment or microprocessors at all, or
- Do what you think is suitable and hope it will be all right.

Neither of the above choices was acceptable, particularly because of the competitive nature of the manufacturing industries where opportunities for improved productivity with assured safety cannot be ignored.

Consequently the electrical engineering communities have worked hard to develop suitable new standards to cover the principles of applying programmable systems to what is known as functional safety. In this module we explain some of the basics of the new standards and show their relevance to machinery safety.

9.2 Objectives

- To outline standard IEC 61508 and its main features
- To explain its relevance to machinery safety
- To explain the concepts of safety integrity levels
- To introduce the recently issued standard IEC 62061 for functional safety of machines
- To provide an introduction to the method of quantified risk assessment and show how SIL targets can be determined.

9.3   Outline of IEC 61508

IEC 61508 is titled: 'Functional Safety of Electrical/Electronic/Programmable Electronic Systems'. It is intended to be a widely applicable 'generic standard' which suggests that it can be applied to many types of safety systems in many types of industry. The term 'generic' implies that we can use a more specific version of its rules to suit a particular type of industry. It has been prepared with this idea in mind, which has led to situations such as the one described by Figure 9.1.

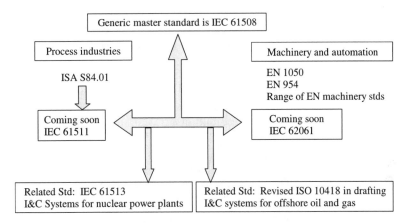

Figure 9.1
Existing and emerging standards for functional safety

This figure indicates that particular industry sectors have developed their own specific interpretations of the 'master standard' best suited to the conditions and practices within their sector. Hence we can see that the process industries have IEC 61511 and the machinery and automation sector has IEC 62061.

The sector standards have not yet emerged from the final draft stages, so the present application of IEC 61508 is largely based on the general guidelines that it provides. In the areas of hardware and software engineering, major companies are following these practices quite rigorously and there are a number of features that are already significant at this stage for machinery applications. The term 'functional safety' also requires some explanation.

9.3.1   Functional safety

'Functional safety' is a concept directed at the safety device or safety system itself. It describes the aspect of safety that is associated with the functioning of any device or system that is intended to provide safety.

> *In order to achieve functional safety of a machine or a plant the safety related protective or control system must function correctly and, when a failure occurs, must behave in a defined manner so that the plant or machine remains in safe state or is brought into a safe state.*

> **From:** *"Functional Safety in the field of industrial automation" by Hartmut von Krosigk. Computing and Control Engineering Journal (UK IEE), Feb. 2000.*

> *Functional safety is that part of the overall safety of a plant that depends on the correct functioning of its safety related systems. (modified from IEC 61508 part 4)*

This description is equally applicable to machinery safety-related control systems and hence the new standards are fully applicable in this area.

9.3.2   Standard IEC 61508

The correct titles of the standard and its parts are shown in Figure 9.2.

Introducing standard IEC 61508

International Electrotechnical Commission

Title:
Functional safety of electrical/electronic/programmable electronic safety-related systems –

All Sections of IEC 61508 Now Published

Part 1: General requirements
Part 2: Requirements for electrical/electronic/programmable electronic systems
Part 3: Software requirements
Part 4: Definitions and abbreviations
Part 5: Examples of methods for the determination of safety integrity levels
Part 6: Guidelines on the application of parts 2, and 3
Part 7: Overview of techniques and measures

Figure 9.2
Title and contents of IEC 61508

This standard is published in seven parts and the role of each part can be summarized from our perspective as follows:

- *Part 1:* Tells us how to manage the overall safety project by using the safety life cycle approach. It uses the safety life cycle as framework for a set of requirements to be carried out at each phase of the project.
- *Part 2:* Defines SIS design requirements and the detailed procedures to be observed in developing, building and testing the equipment.
- *Part 3:* Details the software engineering practices that must be observed for a programmable system to qualify for safety duties. It scales the special engineering requirements against the SILs. This part is largely aimed at developers of operating systems for safety-certified controllers.
- *Part 4:* Provides definitions of terms.
- *Part 5:* Provides advice on methods of determining the SIL requirements from information obtained from hazard studies. Here we find the various methods of quantitative and qualitative analyses including risk graphs.
- *Part 6:* Provides guidance on how to carry out the requirements defined in parts 1, 2 and 3. In particular, this part contains useful sections on how to do the reliability calculations used to evaluate the SIL of a proposed design.
- *Part 7:* Provides references to further reading and techniques used in support of the SIS design work.

9.3.3 IEC safety life cycle

The safety life cycle model provides a framework for project activities supporting a safety system application. It looks like as shown in Figure 9.3.

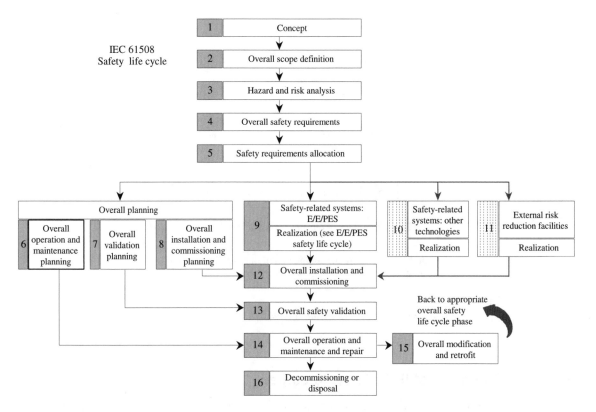

Figure 9.3
The safety life cycle as described in IEC 61508

Any project team seeking to comply with the IEC standard is required to plan and document its project activities along the lines set down in this safety life cycle model (we abbreviate it here as SLC). The standard provides detailed requirements for the tasks to be performed at stage of the SLC. These are set down in paragraphs identified against each box on the SLC diagram. Each box is a reference to a detailed set of clauses defining the requirements of the standard for that activity. The boxes are easy to follow because they are defined in terms of:

- Scope
- Objectives
- Requirements
- Inputs from previous boxes
- Outputs to next boxes.

The fundamental rule is that each box is based on information contained in the preceding box. Just as in any QA system it is important that each stage is complete and agreed before the next stage can be signed off (see Figure 9.4).

Whilst concurrent design is allowed the final result must be that all boxes are aligned and finalized. The SLC scope includes all stages from initial concepts and hazard studies through to operation, maintenance and modification. The standard covers electrical, electronic and programmable electronic systems and lays down standards of engineering and quality assurance for both hardware and software.

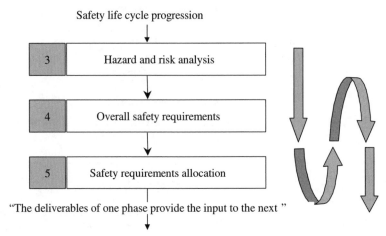

Safety life cycle progression

3	Hazard and risk analysis
4	Overall safety requirements
5	Safety requirements allocation

"The deliverables of one phase provide the input to the next"

It does not require the completion of one activity before starting another: i.e. "a concurrent design approach can be used".

Figure 9.4
Safety life cycle progression allows for interaction between stages

9.3.4   Summary of IEC life cycle stages

Here is a simplified version of the SLC drawn to show briefly what the project cycle entails (see Figure 9.5).

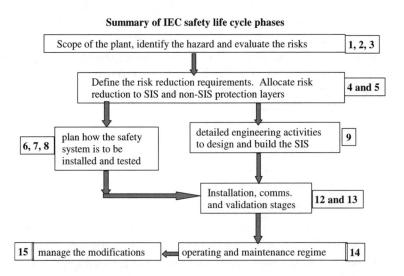

Figure 9.5
Simplified version of the safety life cycle

9.3.5   Key elements of IEC 61508

- *Management of functional safety is just as important as the way we design the functional safety system:* You cannot claim compliance with the standard unless you can demonstrate that the project has been run under a formalized set of management procedures ensuring everyone knows what they are required to do to support the safety systems at the plant.
- *Technical requirements cover hardware, software, testing:* SIS designs must satisfy constraints on architectures and safe failure fractions. Equipment must have essential proven characteristics. This leads to the certification of devices or subsystems by third parties, e.g. TUV SIL-3 certification. Self-certification is also permitted.
- *Documentation is mandatory and must be kept current:* Companies must keep a complete set of records of the SLC activities and show that the SIS as installed is valid for the current plant design and its hazard analysis. The testing regimes must match the SIL performance requirements. The testing records must line up with the reliability analysis.
- *Competence of persons:* Individuals assigned by companies to work on safety systems must be competent to perform the tasks. The standard does not detail the experience of skills needed but other bodies have developed guidelines.

9.3.6   How is this relevant to machinery safety?

The relevance of IEC 61508 is that it provides a common standard for the design of safety-related control system using programmable systems. This will enable the following:

- Better definition of the overall safety performance of protection systems through the use of safety integrity levels (SILs). (Safety integrity levels cover the overall safety performance of a system rather than just the hardware and self-testing features described by a safety category.)
- Machinery protection functions can be specified using SILs. These will enable SIL-rated equipment and SIL compliant designs to be matched to the SIL ratings specified for each application.
- Machinery protection equipment such as safety PLCs and sensors can be designed to defined SILs to match the user's requirements.

9.3.7   Problem with safety categories

We have seen how EN 954 uses the risk assessment information to decide on a safety category. We have also seen how the safety categories define the amount of redundancy and self-testing needed in a safety-related control system. The problem is that the category descriptions and their performance needs are only suitable for hardware-based solutions and do not translate easily to describe programmable systems that combine hardware and software to do their job.

Safety system designers now use the SIL concepts to develop and certify safety systems because these describe the expected performance of the combined hardware and software systems. To fully utilize the advantages of programmable systems in machinery safety we need to be able to determine SIL categories for the safety functions in the machines. So let us examine the concept of SILs.

9.4   Concept of SILs

The SIL rating of a safety function is a measure of the degree of confidence in its overall ability to provide the safety function for nearly all of the time and under nearly all circumstances. We have used the term 'nearly' because no safety system can be 100% effective. The closer the safety system gets to perfection, the higher will be its SIL rating.

The confidence level or SIL rating of the safety system will depend on a number of key factors that can be seen most simply in Figure 9.6.

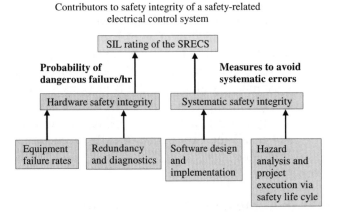

Figure 9.6
Factors contributing to the safety integrity of a safety-related electrical control system (SRECS)

The figure indicates that it is not sufficient to just have reliable hardware.

- The dangerous failure rate of the hardware is very important but the probability of a dangerous failure due to random hardware faults can also be reduced by the familiar arrangements of redundancy that we have seen earlier in the workshop. (Also known as redundant architectures.)
- Safety integrity also depends on the avoidance of systematic errors in specification and design of the safety functions. This has to be achieved by the careful application of SLC activities such as hazard studies, development of the safety requirements specification and the implementation of the design and installation phases. The practice of regular verification of design results followed by final validation of the system through testing is also most important.
- For programmable systems the safety integrity of software has to be assured through the use of operating systems and programing packages that have been developed according to the software SLC procedures laid down in IEC 61508, part 3.
- Application software must be developed under quality assurance procedures giving a high degree of confidence in its ability to perform the specified safety functions.

The IEC 61508 standard provides the necessary procedures and guidelines to meet all these requirements.

9.4.1 How are SIL performance and design features defined for machinery safety?

For hardware reliability, the SIL rating is obtained from the estimated probability of a dangerous failure per hour (PFH). IEC 61508 part 1 Table 5 carries a range of PFH values qualifying for SIL ratings as follows:

IEC 61508 part 1 Table 3: Safety integrity levels
for continuous mode of operation

Safety Integrity level	Continuous Mode of Operation (probability of a dangerous failure per hr)
4	$>10^{-9}$ to $< 10^{-8}$
3	$>10^{-8}$ to $< 10^{-7}$
2	$>10^{-7}$ to $< 10^{-6}$
1	$>10^{-6}$ to $< 10^{-5}$

Probability of a dangerous failure per hr is also known as the dangereous failure rate per hr or the frequency of dangerous failures.

When translated into yearly failure rates it can be seen that SIL-1, for example, lies in the range of one failure per 10 years to one failure per 100 years.

Redundancy and diagnostic features in hardware subsystems (i.e. input sensor subsystems or logic solvers or final elements) are linked to SIL ratings by the following table of constraints:

Safe Failure Fraction	Hardware Fault Tolerance		
	0	**1**	**2**
<60%	Not allowed	SIL-1	SIL-2
>60% to <90%	SIL-1	SIL-2	SIL-3
>90% to <99%	SIL-2	SIL-3	SIL-3
>99%	SIL-3	SIL-3	SIL-3

Note that SIL-4 ratings are not considered appropriate for machinery applications.

These constraints generally mean that redundancy arrangements of 1oo2 are needed for most SIL-2 and-3 systems and sometimes 1oo3 arrangements are needed for SIL-3. The constraints also leave room for reduced redundancy where high safe failure factions can be achieved. This is normally obtained only by the use of a high level of self-testing or diagnostic activity within the safety subsystem. This is commonly achieved in high-performance safety PLCs.

9.4.2 How do the SIL ratings compare with safety categories?

The new IEC 60261 has provided an approximate alignment of SIL ratings with the safety categories as defined by EN 954. This can be seen most easily in the fact that SIL-3 safety systems require a fault tolerance level of at least 1 with a high level of self-testing. The basic details of alignment are as follows based on information from Table 3 in IEC 62061:

Safe Failure Fraction	Hardware Fault Tolerance	Category to EN 954	Maximum SIL Claim
	0	B	Not valid as a SIL
<60%	0	1	1
>60% to <90%	0	2	1
<60%	1	3	1
>60% to <90%	1	3	2
>60% to <90%	>1	4	3
>90%	1	4	3

The indications from this table are that it will be possible to use most category-rated hardware solutions to existing safety system applications alongside SIL-rated equipment. The new methods of classifying programmable systems will therefore be easy to align with existing category-rated solutions.

9.5 How can we determine the required SIL for a safety function?

The following outline description of SIL determination for machinery SRECS is based on suggested methods set down in the draft version of IEC 62061 'Safety of machinery: functional safety of electrical/electronic/programmable electronic control systems for machinery'. The description is a very shortened and approximate version of a comprehensive estimating guide that has been drafted for approval in the standard and it is still subject to review before the standard is released.

For machinery systems the SIL rating of an SRECS is determined on the assumption that the controls are working in a continuous or 'high demand' mode of operation. This leads to the simple assumption that the machine will be in a dangerous condition whenever the SREC fails. The frequency and scale of the potential accident will then depend on the likelihood of a person being in the danger zone and being hurt when the safety system has failed.

The basic principle employed to work out the required SIL begins with an estimating procedure to predict the scale and frequency of injury that would be expected if the regular machine controls were used without any special safety provisions (see Figure 9.7).

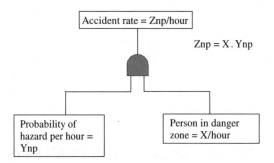

Figure 9.7
Initial prediction of the accident rate without a safety system

From the predicted accident scenarios and frequencies a review procedure leads to a decision on the required 'improvement factor' in the worst-case accident rate. Presuming that an SRECS is to be used to provide the desired improvement (i.e. reduction) in accident rate the new rate will be a function of the fail to danger rate of the SRECS as shown in Figure 9.8.

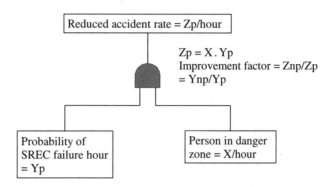

Figure 9.8
Revised prediction of the accident rate with the safety system in place

The required improvement factor is used to determine the required SIL of the SRECS using the following table:

SIL	Risk Improvement Factor	PFH of the SRECS
1	10 – 100	10^{-6} to 1^{-5}
2	100 – 1000	10^{-7} to 10^{-6}
3	1000 – 10000	10^{-8} to 10^{-7}

So, for example, if we call for a risk improvement factor of 500 we shall require a SIL-2-rated SRECS. The design of the system will have to be good enough to ensure that its probability of a dangerous failure per hour is at least $10^{-6.}$ This is the equivalent of an MTBF danger of 10^{6} h or approximately 100 years.

Hence the performance level of safety instrumentation needed to meet the SIL is divided into a small number of categories or grades. The SIL ratings provide risk improvement factors to reduce the accident rate on the machine and they also determine the engineering measures needed to deliver the prescribed level of safety integrity.

A SIL-1 system is not as reliable in the role of providing risk reduction as SIL-2; a SIL-3 is even more reliable. Once we have the SIL we know what quality, complexity and cost we are going to have to provide to meet the required safety performance. For a further review of this topic please refer to Section 9.6.

9.6 Some implications of IEC 61508 for machinery systems

1. This standard is the first international standard that sets out a complete management procedure and design requirements for overall safety control systems. It is currently being extended to machinery control as IEC 62061 where it is likely to complement existing standards that do not cover programmable equipment.

2. Control systems and PLCs serving in safety-related applications may be required in the future to be in conformance with the requirements laid down in IEC 61508. Machinery safety requirements will eventually need to conform to IEC 62061 using the principles laid down in IEC 61508.

3. All forms of control systems with any potential safety implications could be subject to evaluation or audit in terms of IEC 61508.

4. Design and hardware/software engineering of any safety-related control system is to be evaluated and matched to required SILs. This places responsibilities on the end user and designers to demonstrate through reliability analysis and documentation that their SRECS meet the target performance.

5. Software engineering procedures and software quality assurance are mandatory requirements for a PES in safety applications. The standard provides the basis for certification of software packages by authorities such as TUV.

6. Responsibilities of users and vendors are clearly defined:

   - The user must define his requirements in terms of functional safety.
   - The vendor must show how his solution meets the requirements in terms of the user's specific requirements.
   - The user's responsibilities for operation, maintenance and change control are defined as part of the conformance.

What will be the role of IEC 62061?

The new IEC standard for functional safety systems in machinery is designated IEC 62061: Safety of machinery – functional safety of electrical, electronic and programmable control systems.

Some key points to note about IEC 62061:

- As a sector standard it relates directly to IEC 61508 and hence incorporates the principles that have been accepted internationally for PES-based functional safety systems.
- The specification and design of safety systems is based on SILs. SILs define hardware, software and management of the design and its implementation as essential components of safety integrity.
- It is based on quantified risk assessment principles as laid down in EN 1050 (now known as ISO 14121).
- It follows the principles of design for machinery safety as laid down in EN 292 (now known as Pr EN ISO/FDIS 12100).
- Allows hardware subsystems based on EN 954 and its safety categories to be integrated into the SIL-based safety systems.

- It covers the SLC for machinery controls from safety requirements definition to final design and validation.
- It is expected to align with other existing standards and relate to them as shown in Figure 9.9.

Effectively this standard allows us to specify the machinery safety controls in terms of SILs or alternatively we may use the existing safety categories where these would be valid. This in turn allows us to make use of SIL-rated devices on the market to build up a safety system for a machine using safety-certified PLCs, smart sensors and networks all within the framework of an accepted international standard for functional safety, namely IEC 61508.

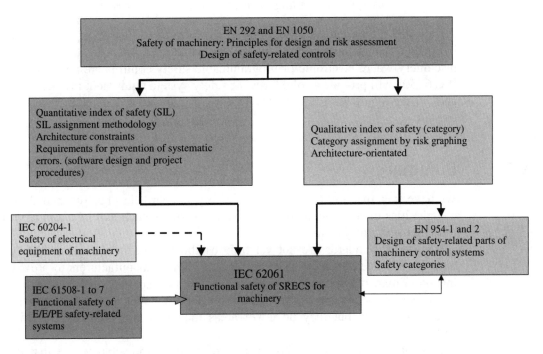

Figure 9.9
Relationship of IEC 62061 to other relevant safety standards

The main purpose of IEC 62061 is claimed to be 'to facilitate the specification of the performance of electrical control systems in relation to the significant hazards of particular machines within machine-specific standards'.

This suggests that it will help provide a consistent design reference for persons writing safety standards for particular types of machines. It also looks as it will be a useful guide to good practices in any automation safety project. Let us take a look at some features and how they may affect machinery safety design work in the future.

- The method of estimating risk and defining SILs
- How safety categories compare with SILS
- The method of structuring a safety function.

This new standard for functional safety in machinery will provide an SLC plan for the development and application of SRECS. The drafts seen to date include:

- Project life cycle procedures.
- A method for structuring each safety function into subsystems that is typical for the way in which a number of safety functions are shared within a complex machine control system. This deals with the problem of specifying the SIL requirements for multiple safety functions.
- Design requirements for subsystems to meet functional and integrity requirements.
- SIL tables and equivalence tables for EN 954 categories.
- Appendices to assist with the determination of required SILs. These include a numerical method for allocating risk factors to frequencies and accident consequences leading to an SIL requirement for each function.

The role of this standard appears likely to be found in support of individual applications by makers of manufacturing automation plant and the end users of any complex machinery where electronic or programmable safety controls are being deployed. Whilst IEC 61508 will provide a solid basis for safety system equipment manufacturers to design and certify, equipment IEC 62061 is clearly intended for machinery builders and end users.

9.7   Summary

We have seen that a new generation of safety standards has been launched to support the complexities of new technologies for safety systems such as PLCs and microprocessor-based sensing subsystems.

These new standards are not yet fully established in the machinery safety sector but have already been used extensively by safety device manufacturers to provide assurance that programmable system products are suitable for machinery applications.

The new standards generally do not conflict with existing methods and standards for machinery safety but they do cover areas that were not previously considered in the earlier standards.

This places end users in the position of being responsible to see that their projects conform to the new standards and raises the likelihood that regulators will expect to see conformance to the new standards.

The task for the end user is for technical staff to become familiar with the basic requirements of the standards and for plant management to become aware of the need for appropriate management systems to be in place.

9.8   Conclusion

All of this looks quite daunting for the project engineer faced with an automation task for perhaps integrating several machines into an assembly line. However the process of risk assessment has not been changed by this standard; it remains true to the detailed safety principles set down in EN 292 and the risk assessment principles detailed in E 1050. So there is perhaps no real increase in burden here.

The advantages of using SIL ratings for the safety function is that the standards for programmable systems can then be directly applied and the safety system equipment manufacturers will offer products that are compliant with particular SIL ratings based on IEC 61508.

It is also important to note that it has just been announced by the European standards organizations that the EN 954 standard is to be replaced by a new standard to be numbered ISO 13849-1. It is believed that this standard will follow the route of IEC 62061 in specifying risk reduction performance in terms of 'performance levels' or PLs in same way as SILs. At this stage it is not clear whether SILs and PLs are effectively the same. (Reference source: PILZ Newsletter, April 2003.)

Looking at the present situation we can conclude with the following points:

- The established methods of defining safety requirements by category remain suitable for simple hard-wired safety applications but have become unsuitable for PES applications.
- Safety categories do not adequately describe the type of safety system solutions now available in PES devices. There is risk that many existing machinery builders are not in compliance with good practices for PES-based safety solutions.
- Safety-certified PLCs and networks have become established technology for automation in machinery.
- The lack of a suitable standard has been overcome so far by showing equivalent performance can be obtained by systems compliant with IEC 61508. This has been an inefficient and temporary way to deal with the changing technologies of safety.
- The new sector standard IEC 62061 or the planned replacement for EN 954, ISO 13849-1 (Safety of machinery, Safety-related parts of control systems, General principles for design) will hopefully assist designers to specify an SIL rating for each safety function in a project. However this will require a cultural shift from qualitative risk assessment to greater use of quantitative risk assessment.
- Using SIL-based design methods in the new standards will help to unify the design practices for machinery safety with those in other industries. The new standards will present internationally accepted practices for ensuring safety in the high-performance machines and control system of the future.

References

[1] Siemens Safety Integrated Application Manual. www. Siemens.de/safety.
[2] PILZ: Guide to machinery safety, 6th edn, 1999. www.pilz.com.
[3] PILZ: Guide to programmable safety systems, Vol. 2, 1st edn.
[4] Guardmaster: product guide and safety navigator.
[5] IEC 61508-1-7 Functional safety of electrical/electronic/programmable electronic safety-related systems. International Electrotechnical Commission 3, rue de Varembé1211 Geneva 20, Switzerland. www.iec.ch.
[6] IEC 61511 parts 1–3: Functional safety: Safety instrumented systems for the process industry sector.
[7] Committee Draft for IEC 62061: Safety of machinery – Functional safety of electrical, electronic and programmable control systems, version 44/380/CD, 1/05/2002.7.
[8] IDC Technologies: training workshop manual: Practical safety instrumentation and emergency shutdown systems. Ver 7.0, 2002. www.idc-online.com.

Appendix: Notes on the method for the determination of SILs for a machinery safety application

These notes are appended to provide an example of how quantified risk assessment can be used to determine the required safety integrity level for a safety function. This method was introduced in the text, but a more complete example has been provided here to assist in training work.

SIL requirements are defined by evaluating the required fail to danger rate of the safety-related system that will deliver an acceptable risk. The basis of acceptable risk is the accident rate that is similar to that experienced in everyday life. An example of this can be seen in Figure 9.10 figures based on diagrams published by the UK Health and Safety Executive.

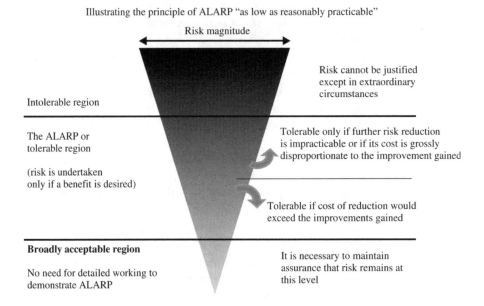

Figure 9.10
Alarp diagram

The Alarp principle states that risk should be reduced to as low as reasonably practicable provided it is below the intolerable level. As risk is further reduced it enters the broadly acceptable region where no additional safety measures are needed.

In Figure 9.11 the broadly acceptable risk levels are indicated for the United Kingdom based on consensus information. Notice that this presents target accident rates for different classes of consequence. This means, for example, that if the probability of a fatal accident for an individual is reduced to below 10^{-6} per year it will be no more risky than everyday life for a member of the public. So we have numerical targets for the risk levels that we would like to achieve. The same principles apply to target rates for accidents with serious consequences and for minor consequences.

Using quantitative risk targets makes it possible to draw up reliability performance targets for safety-related control systems. An early draft version of IEC 62061 proposed the following method to arrive at the required SIL rating for a safety function. It serves here as a typical example of a quantitative method.

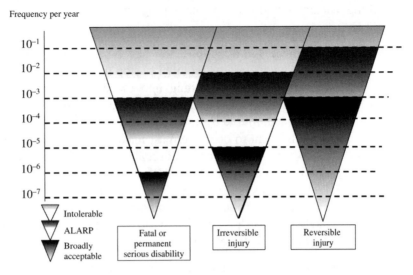

Figure 9.11
Alarp diagrams for classes of consequence

In the first step, the accident rate is derived by combining the frequency of a person becoming exposed to the danger with the frequency of a dangerous failure of the safety function. Figure 9.12 is a fault tree model of the potential accident situation.

The fault tree model indicates that an accident occurs if a person becomes directly exposed to danger at the same time that the safety function designed to protect against the accident has failed to a dangerous state.

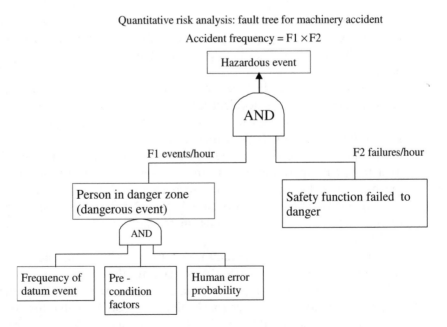

Figure 9.12
Quantitative risk analysis: fault tree for machinery accident

As an example: Suppose that an operator has to place a workpiece in a multi-function milling machine running an automatic program.

- The datum event is that he reaches across the machining table to unlock the workpiece. Let us suppose this has to be four times every hour per 8-h working shift. The operator is required to stop the machine and allow it time to cease all movement before reaching into the danger zone. An accident occurs if he reaches into the danger zone before the operating cycle has finished and he gets caught and injured by a moving cutting tool.
- For this event to occur he must make a human error of repetition by forgetting to stop the machine or by not realizing the machine is moving. The probability of this has been taken to be 0.05.
- Preconditions such as machine defects or misalignment of the workpiece modify the probability of human error. In this example we can take the likelihood as 'probable' but not 'frequent'. The draft standard suggests a probability factor of 0.01.

The frequency of the datum event is calculated for this machine based on the operator's involvement time with the machine. In this case we estimate it to average 4 events per hour of the operator's time at work.

For estimating preconditions we need to consider what factors would be likely to cause the error. The precondition is considered to be part of the sequence of events that leads to the potential accident and it provides a probability factor for the event. The contributing factors will include:

- *Human factors:*   Time pressure to produce more output, ignoring of stated procedures and loss of concentration.
- *Environment:*   High noise levels, poor access to the machine, poor lighting.
- *Machine condition:*   Poor maintenance, inadequate or poorly fitted guards.
- *Machine operation:*   Inadequate stopping performance, operation in wrong cycle.

This is the most difficult aspect of the quantitative risk estimation and it requires that an experienced person should examine the intended use of the machine and review all the factors that may influence the rate at which the operator will make an error.

The fault tree model in Figure 9.13 tells us that the event frequency for the operator being in exposed to the accident at any time is $4.0 \times 0.01 \times 0.05 = 2.0 \times 10^{-3}$ per hour. This predicts an approximate accident rate of four events per year (2000 h per year) for this worker if no safety devices are fitted.

The operator is to be protected by a safety function, which in this case is proposed as a light-curtain screen across the approach to the cutting table that will trip the machine to an E-stop if the infrared beams are interrupted. For the moment we will assume this function has not been SIL-rated but is part of the basic control system for the machine. We will apply an estimated failure rate of 1×10^{-4} faults per hour as a trial value.

Applying this figure to the fault tree delivers an accident rate of 1×10^{-7} per hour or a risk rate of approximately 1.0×10^{-3} per year for the operator.

The next step is to compare this rate with the risk criteria guidelines for the likely category of injury for the operator. Notice, in Figure 9.14 that the Alarp diagram presents target accident rates for different levels of consequence. So we have numerical targets for the risk levels that we would like to achieve. If we assume this is an irreversible injury we see that the target risk rate should be at least as low as 10^{-5} per year.

Example: Fault tree for milling machine accident. Step 1

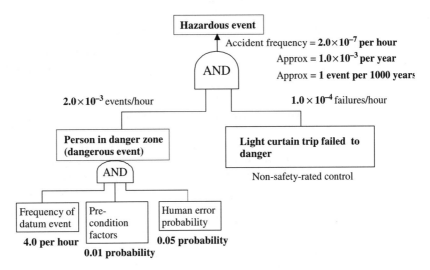

Figure 9.13
Trial solution with non-safety-rated control

SIL assignment procedure

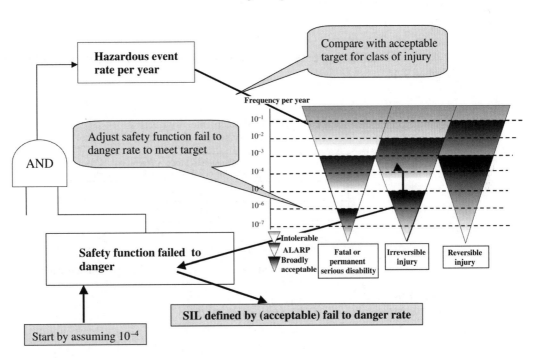

Figure 9.14
Adjustment of safety function SIL to meet target event rate

To achieve this we need to improve the safety function failure rate so that F1 × F2 < 10^{-5} per year or 10^{-9} per hour (F1 = unprotected accident rate, F2 = target failure rate of protection system) as shown in Figure 9.15.

$$F2 = \frac{10^{-9}}{(2.0 \times 10^{-3})} = 5 \times 10^{-7}$$

Example: fault tree for miling machine accident, step 3

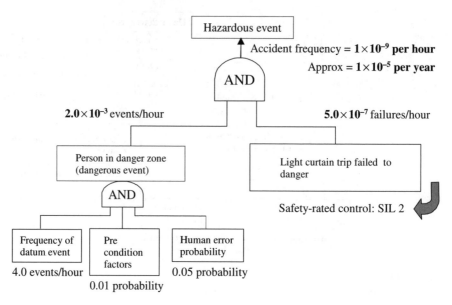

Figure 9.15
Revised risk analysis with SIL 2 rated safety control

The SIL table indicates that this requires an SIL-2-rated safety function (see Figure 9.16).

IEC 62061 safety integrity levels
for continuous mode of operation

Safety Integrity Level	Continuous Mode of Operation (probability of a dangerous failure per hour)
3	$>10^{-8}$ to $<10^{-7}$
2	$>10^{-7}$ to $<10^{-6}$
1	$>10^{-6}$ to $<10^{-5}$

SIL 2 5×10^{-7} **failures/hour**

Figure 9.16
Using the SIL table to convert target failure rate to SIL

Discussion

A similar quantified method of SIL determination has been included in the final version of IEC 62061. The standards committee has recognized that fault tree methods require specialized skills and proposed a series of 'fill in the blanks' forms that would assist users to follow a systematic procedure to arrive at the SIL rating (if any is needed) for each safety function. However, as can be seen from the fault tree example the result of the risk estimation process depends heavily on the estimates of the factors predicting the exposure and error rates of the persons close to the machine.

In calling for a SIL-2 solution for the milling machine example we would expect to have a category 3 rated light curtain with a safe failure fraction in the range 60–90%.

Using the EN 954 guidance chart we found the category would probably be 3. No real disagreement here but the benefit of defining the solution as a SIL rating is that all the important features of a SIL-2 safety system are spelt out in the IEC 61508 (and in the upcoming IEC 62061) standard for programmable systems. It tells us how to build a SIL-2 system in hardware and software and we can buy SIL-2-rated equipment certified by a testing/assessment authority.

The other advantage of the SIL determination method is that the risk assessment model can be set out with quantified probability values. This allows the circumstances of the safety system to be carefully modeled and recorded for further evaluation or auditing.

Appendix A

References and sources of information on machinery safety

References and information sources used in preparing the manual

Ref. No.	Title/Subject	Origin
1	Out of Control: Why control systems go wrong and how to prevent failure. ISBN 0-7176-0847-6	UK Health and Safety Executive. HSE Books. www.hse.gov.uk
2	HB142-1999: A basic guide to managing risk using the Australian and New Zealand risk management standard	Standards Association of Australia. PO Box 1055, Strathfield NSW 2135. www.Standards.com.au
3	Tolerable risk guidelines	Edward M Marzal: Principal Engineer, Exida.com. www.exida.com
4	HAZOP and HAZAN by Trevor Kletz, 2nd edn, 1986	I Chem. Eng Rugby, UK
5	Guide to machinery safety, 6th edn, Feb. 1999 and Interactive safety guide, vers. 4.0 CD	PILZ Automation Technology, Medlicott Close, Corby, Northants, UK. www.pilzsupport.co.uk
6	Specifier's guide to machine safety and safety source CD-ROM	Banner Engineering Corp, Minneapolis, MN 55441, USA. www.baneng.com
7	Safe machinery with opto-electronic protection and Safexpert 3.1 Demo CD	Sick AG: Safety Systems, PO Box 310 D-79177 Waldkirch, Germany. www.sick.de
8	A-B-Guardmaster safety products application manual and Safety Navigator CD Rockwell Automation website for guideline features on machinery safety	Rockwell Automation, A-B Safety Products. www.ab.com

Ref. No.	Title/Subject	Origin

9	Safety integrated applications manual	Siemens
		Automation and Drives Group,
		Low-Voltage Switchgear Business
		   Division, Postfach 3240, D-91050
		   Erlangen, Germany.
		www.siemens.de/safety

10	IEC 61822: Hazard and operability studies (HAZOP studies) – Application Guide, 1st edition 2001–05	International Electro-Technical Commission, Geneva, Switzerland. www.iec.ch
11	Guidelines for process hazard analysis: Hazards identification and risk analysis by Nigel Hyatt	Dyadem International Ltd, Toronto, Canada. www.dyadem.com
12	HAZOP guide to best practice by Frank Crawley, Malcom Preston and Brian Tyler. ISBN 0-85295-427-1	Published by: Inst. of Chemical Engineers, Rugby, UK. www.icheme.org.uk
13	EN 292-1: Safety of machinery – basic concepts, general principles for design – Part 1, Basic terminology, methodology EN 292-2 – Part 2 – Technical principles and specifications EN 1050: Safety of machinery – Principles for risk assessment EN 954-1 1997 Safety of machinery – Safety - related parts of control systems – Part 1. General principles for design	CEN, B-1050 Brussels British Standards Institution (for BS EN 292, etc.) www.bsiglobal.com
14	IEC 61508: Functional safety of E/E/PES systems, Parts 1–7	International Electro-Technical Commission, Geneva, Switzerland. www.iec.ch
15	Draft IEC 62061: Safety of machinery – Functional safety of electrical, electronic, programmable control systems for machinery. IEC/TC 44 vers. 44/380/CD	
16	IEC 602041-1 1997 A1-1999: Safety of machinery – Electrical equipment of machines. Part 1, General requirements	
17	Safe use of work equipment. PUWER 1998, Approved code of practice and guidance	The Health and Safety Executive, HSE Books, PO Box 1999, Sudbury, Suffolk CO10 2WA, UK. www.hsebooks.co.uk or via www.hse.gov.uk

18	HSE free publications:
	  5 steps to risk assessment
	  A guide to risk assessment requirements
	  Supplying new machinery
	  Buying new machinery
	  Reducing risks, protecting people
	  Simple guide to the provision and use of work equipment regulations 1998	

19	Europa website (info on machinery and other directives + listings of harmonized standards referenced by individual directives). http://europa.eu.int/comm Machinery section: http://europa.eu.int/comm/enterprise/mechan_equipment/machinery/index.htm	

Suggested references and further sources

Item No.	Title/Subject	Origin/Author
1	IEC 60812: Analysis techniques for system reliability – Procedure for failure modes and effects analysis (FMEA)	www.iec.ch
2	Reliability maintainability and risk, 6th edn, 2001	Butterworth-Heinemann/ Dr David J Smith
3	MIL-STD-1629A: Procedures for performing a failure modes, effects and criticality analysis	US Dept of Defense (1980)
4	PILZ newsletter	www.pilzsupport.co.uk
5	AMT Technical report, ANSI B11.TR3 – 2000. (Details of suggested risk assessment procedures compatible with US laws but similar in nature to EN 1050.)	Association of Manufacturing Technology, USA
6	Safe use of power presses. PUWER 1998 as applied to power presses. Approved code of practice and guidance L112	The Health and Safety Executive, HSE Books, PO Box 1999, Sudbury, Suffolk CO10 2WA, UK. www.hsebooks.co.uk or via www.hse.gov.uk
7	Safe use of woodworking machinery. PUWER 1998 as applied to woodworking. Approved code of practice and guidance L114	
8	Safe use of lifting equipment. Lifting Operations and Lifting Equipment Regulations 1998. Approved code of practice and guidance L113	

Appendix B

Glossary

ANSI American National Standards Institute. Source of the B11 series standard for machinery safety.

Availability The probability that an item of equipment or a control system will perform its intended task (see safety availability).

Basic control system Generic term used to describe any control system equipment provided for the normal operation of a plant or machine. A basic control system may or may not include safety functions but if it does it these will be known as the safety-related parts.

Blanking Feature of a safety light curtain system that allows the optical sensors to ignore fixed or moving objects known to be safe in the defined area of protection.

CASS Conformity assessment of safety-related systems. Refers to the developing methods for assessment of project execution, equipment design as well as functional safety management capabilities. In the UK, accredited certification bodies offer CASS assessment services to industry.

Cause and effect diagram A matrix drawing showing the functional process safety interlocks between inputs and outputs of a safety system (see also FLD).

CE "Conformité Européenne" Conformity symbol showing claimed compliance for a machine or product to the relevant EU directives.

Common-cause failure Failure as a result of one or more events, originating from the same external or internal conditions, causing coincident failures of two or more separate channels in a multiple channel system (see also systematic failures).

Control reliability US term requiring safety controls (SRECS) to be fault-tolerant but also requiring that they detect the fault and prevent the next operation of the machine from starting until the fault has been corrected.

Coverage factor See diagnostic coverage.

Covert failure A hidden defect in a safety system that is not detected by the incorporated test.

De-energized safe condition In this context: the electrical or pneumatic devices that can shutdown the guarded process are energized during the normal (safe) operating situation. If an unsafe condition arises, the spring-loaded device will revert to the off or 'shelf' condition.

Diagnostic coverage   A measure of effectiveness of the self-diagnostics of a safety-related control device or function, which makes it possible that a system successfully detects a specific type of hardware or software fault. IEC 61508 defines diagnostic coverage as 'fractional decrease in the probability of dangerous hardware failures resulting from the operation of the automatic diagnostic tests'.

Diagnostic coverage factor (also known as C-factor)   The C-factor comprises the percentage of failures in modules, software, external wiring, internal wiring, cables, interconnections and other functions that are detected by the built-in test functions, or by a suitable test program. It can be expressed in a probability or in a factor that is always smaller than 1 (e.g. C = 0.95) or as a percentage (e.g. 95%).

E/E/PES (Electrical/electronic/programmable electronic system)   Formal term used in IEC 61508 for a 'System for control, protection or monitoring based on one or more electrical/electronic programmable electronic (E/E/PE) devices, including all elements of the system such as power supplies, sensors and other input devices, data highways and other communication paths, and actuators and other output devices'.

EMC   Electromagnetic compatibility.

Emergency shutdown   In machinery practice this is a trip or emergency stop. In more complex machines or processes it refers to a controlled sequence of actions to make the machine safe as quickly as possible.

EMI   Electromagnetic interference.

Fail-safe   A control system that, after one or multiple failures, lapses into a predictable safe condition.

Fault tolerance   IEC definition: 'ability of a functional unit to perform a required function in the presence of faults or errors'. Fault tolerance levels (e.g. 1 or 2, etc.) define the number of such faults that can be tolerated before the system fails to perform its safety function.

FLD   Functional logic diagram. A graphical representation of the system functions, showing the logic gates and timers as well as the logic signal interconnections.

FMEA   Failure mode and effect analysis. Also FMECA: Criticality analysis. Applies when FMEA is extended to identify dangerous or critical failure modes. Also FMEDA: Diagnostic analysis term used for identifying failure modes detected or not detected by diagnostics and for calculating how diagnostics reduce the incidence of dangerous failure conditions (see diagnostic coverage).

Force-guided contacts   Relay or switching device contacts that are mechanically linked such that they always move together. Hence the status of one contact can be used to check the status of the device and its other contacts.

Functional safety   Part of the overall safety relating to the machine control system, which depends on the correct functioning of the safety-related systems.

HMI   Human to machine interface or 'operator interface', usually a computer screen to present the actual process and system status.

Hazard analysis   Term applied to the evaluation of probabilities and consequences of a recognized hazard.

Hazop   Term applied to the structured and systematic examination of a process or system of parts to find possible hazards and operability problems. Hazop is one of a number of recognized hazard study methods.

High demand mode   Where the frequency of demands for operation made on a safety-related system is greater than one per year or greater than twice the proof-check frequency. High demand or continuous mode covers those safety-related systems that implement continuous control to maintain functional safety. These apply to most machinery applications (refer to IEC 61508).

IEC   International Electrotechnical Commission. Based in Geneva. Develops a vast range of internationally supported standards (see Appendix A).

Inherently fail-safe   A particular designed dynamic logic principle that achieves the fail-safe property, from the principle itself and not from additional components or test circuits.

Logic solver   E/E/PES components or subsystems that execute the application logic. A relay system, solid-state logic, pneumatic logic or a PLC can perform this role.

Low demand mode   Applies to a safety control system when the frequency of demands for operation is not greater than twice the proof-test frequency (refer IEC 61511-1).

MTBF   Mean time between failures. This term is normally applied to serviceable equipment, typically sensors, relays, contactors or PLCs. Hence normally used in reliability calculations.

MTTF   Mean time to fail. This term is normally applied to disposable single-life components such as relays or resistors, which are replaced when they fail. Numerically the same as MTBF when calculating reliability of a safety system.

MTTR   Mean time to repair. The mean time between the occurrence of a failure and the return to normal failure-free operation after a corrective action. This time also includes the time required for failure detection, failure search and restarting the system.

Nuisance failure   Describes the trip on non-starting of a machine due to a failure of its safety-related control system rather than because of a hazard.

Overt faults   Faults that are classified as announced, detected, revealed, etc. Opposite of 'covert fault'.

PES   Programmable electronic system. The term includes PLCs as well as any senor or actuator using a programmable device such as a light curtain controller or a communication network.

PFD$_{avg}$ Probability of failure on demand. The average probability of failure to perform the design function on demand (for a low demand mode of operation). This figure is calculated from PFH taking into account the frequency of proof testing (see IEC 61508, part 6).

PFH   The probability of a dangerous failure per hour (for a high demand or continuous mode of operation).

PLC   Programmable logic controller.

Positive-guided contacts   Applies to any actuated switch. Defined in IEC 60204-1 as 'positive opening operation (of a contact element). The achievement of contact separation as the direct result of a specified movement of the switch actuator through non-resilient members (e.g. not dependent upon springs) (see also force-guided contacts).

Proof test   Periodic test performed to detect failures in a safety-related system so that, if necessary, the system can be restored to an 'as new' condition or as close as practical to this condition. Also termed 'trip testing'.

PSDI   Presence sensing device initiation. When an operator moves out of a danger area the machine is automatically started and completes a cycle. Double-break mode is when the first part is removed by the operator who then inserts a new part after which the machine starts. Hence presence detectors require a 'double break' to start the machine. Single-break mode is used when the part is automatically removed. Safety standard defines special measures in the safety systems before allowing this feature.

Redundancy (identical and diverse)   Identical redundancy involves the use of elements identical in design, construction and in function with the objective to make the system more robust for self-revealing failures. 'Diverse redundancy' uses non-identical elements and provides a greater degree of protection against the potential for common cause faults. It can apply to hardware as well as to software.

Reliability   The probability that no functional failure has occurred in a system during a given period of time.

Reliability block diagram   The reliability block diagram can be thought of as a flow diagram from the input of the system to the output of the system. Each element of the system is a block in the reliability block diagram and the blocks are placed in relation to the SIS architecture to indicate that a path from the input to the output is broken if one (or more) of the elements fail.

Revealed failure   A failure in a system that is detected by the systems self-diagnostics.

Safety availability   Probability that an SIS is able to perform its designated safety service when the process is operating. For a low demand rate system the average probability of failure on demand (PFD_{avg}) is the preferred term ($PFD_{avg} = 1 -$ safety availability). For a high demand rate system (e.g. machinery guarding applications) the preferred term is probability of a dangerous failure per hour (PFH).

Safety instrumented system (SIS)   Term used in process control safety and in IEC 61508 for a system composed of sensors, logic solvers, and final control elements for the purpose of taking the process to a safe state when predetermined conditions are violated. Other terms commonly used include emergency shutdown system (ESD, ESS), safety shutdown system (SSD) and safety interlock system. In machinery the equivalent term is safety-related electrical control system (SRECS).

Safety life cycle   Necessary activities involved in the implementation of safety-related systems, occurring during a period of time that starts at the concept phase of a project and finishes when all the E/E/PE safety-related systems, other technology safety-related systems and external risk reduction facilities are no longer available for use.

SCADA   Supervisory control and data acquisition. This term is most commonly applied to PC-based equipment interfaced to plant via PLCs or input output devices.

SER   Sequence of events recorder, based on real-time state changes of events in the system.

SIL   Safety integrity level defining a level of confidence in the risk reduction capabilities of a safety-related device or control system. For hardware aspects this is defined by failure probabilities arranged in order of magnitude. For software and design aspects the SIL relates to the measures taken to ensure quality and freedom from systematic errors in the design. In practice the SIL range is from 1 to 4 as defined in IEC 61508-1.

Solid-state logic   A term used to describe circuits whose functionality depends upon the interconnection of electronic components as semiconductors, resistors, capacitors, magnetic cores, etc. and which does not depend on programmable electronics.

SRECS   Safety-related electrical control system. Used in machinery safety systems to describe any control system that performs a role in reducing risk (see safety instrumented system).

Systematic failures   Failures occurring in identical parts of a (redundant) system due to similar circumstances. History shows that errors in specification, engineering, software and environmental factors, such as electrical interference or maintenance errors must be considered as potential causes of systematic errors or failures. The chances of such failures are reduced by verification and validation of the design and manufacturing stages and by applying structured design methods to software engineering.

TMR   Triple modular redundancy. An architecture for SIS logic solvers to achieve fault tolerance by a 2 out of 3 voting configuration using identical redundant modules.

Trip   A shutdown of the process or machinery by a safety system.

TÜV   Technische Üeberwachungs Verein. A testing laboratory in Germany that certifies safety of equipment in terms of compliance with international standards or German national standards.

Unrevealed failure (undetected failure)   A failure that remains undetected (see also covert failure). These types of failures can be 'safe undetected failures' or 'dangerous undetected failures'. The latter will degrade the safety system or cause it to fail dangerously depending on its level of fault tolerance.

Appendix C

Notes on tolerable risk

This section supports material in Chapter 3.

C.1 Establishing tolerable risk criteria

Is the machine safe? This is the question that always faces a design team during risk assessment. There are both quantitative and qualitative approaches to answering this question.

Quantitative methods compare a predicted accident rate with the recorded accident rate for a given class of consequence. Examples seen in practice include:

- Fatal accident rate (FAR) = Number of fatalities per 10^8 exposed worker hours
- Fatalities or injuries per 100 000 workers in an industry sector
- Probable loss of life (PLL) = Probability of a loss of life due to the hazardous event
- Individual risk (IR) = The frequency at which an individual may be expected to sustain a given level of harm from the realization of specified hazards.

The design team can establish an approximate risk target by comparing the prediction with the average rates achieved in their industry sector. This does not remove the obligation to be better than the average and it will still require careful interpretation.

The UK HSE has published material and reports on the subject of establishing acceptable risk levels and in particular they have available a general guidance publication called 'Reducing Risks, Protecing People. HSE's Decision-Making Process'. This is available for download from www.hse.gov.uk via the search facility. The publication details policy on how tolerable or acceptable risks can be established and it carries some very useful tables of data from accident records. The policy on risks is founded on the 'as low as reasonably practicable (Alarp)' principle we have outlined in the manual.

For an example of how individual risk target values can be used to determine the risk reduction needs for a safety system please refer to the notes on SIL determination at the end of Chapter 9 of this manual. There we have shown a composite Alarp diagram including upper and lower tolerable risk limits for major and minor injuries as well as fatalities. The boundaries are based on figures indicated by general accident statistics and are not intended to be in any way authoritative.

There are also some useful guidelines from a published paper by Edward Marzal at www.exida.com called 'Tolerable Risk Guidelines'. Marzal examined risk management guidelines in several countries and found that some countries have government-based risk tolerance criteria but most countries do not have them. In general the risk criteria are defined by individual companies.

The evidence from government material examined by Marzal included the following tolerable individual risk values for death based on the Alarp diagram with its tolerable risk zone.

- Lower Alarp boundary for a worker in UK: 1×10^{-5}
- Lower Alarp boundary for public in UK: 1×10^{-6}
- Lower Alarp boundary for public in Netherlands: 1×10^{-8}
- Upper Alarp boundary for a worker in UK: 1×10^{-3}
- Upper Alarp boundary for public in UK: 1×10^{-4}
- Upper Alarp boundary for public in Hong Kong: 1×10^{-5}
- Upper Alarp boundary for public in Netherlands and in NSW, Australia: 1×10^{-6}.

To illustrate the meaning of this scale: Assume the IR for worst-case risk to a member of the public in UK is 1×10^{-4}. Take the example of an accident with a 10% chance of causing a fatality. Calculate the highest event frequency that would be considered tolerable. We start by assuming this is the only hazard to which a worker is exposed.

$$F_{max} = \frac{IR}{C} = \frac{(1 \times 10^{-4})}{0.1} = 10^{-3} \text{ events/year}$$

That is it should not be tolerated if its event frequency is higher than 1 in 1000 years.

If the worker has to face several such hazards during his day at work then all the risk event frequencies should be decreased to keep the total risk inside the tolerable zone. On the same basis, if the single event were to cause 10 fatalities this criteria would require the target frequency for that event to be below 1 in 100 000 years.

Edward Marzal points out that the USA is specifically opposed to setting tolerable risk guidelines arguing that they are open to misapplication due to uncertainty about the nature of risks and the population numbers exposed to each risk amongst other factors. He argues that the USA achieves a very good safety record and this is attributed to 'the flexibility to apply capital where it will produce the most benefit and the unrestricted ability of the free market to determine third-party liability costs'.

Marzal concludes that many companies are now finding that where a financial basis for the risk reduction project is calculated the results always justify the greatest amount of risk reduction. For example, a 1995 report by Mudan found that 'risk due to third-party liability of personnel injury is insignificant when compared to other losses such as property damage and business interruption'.

C.2   Using injury statistics as a guide to tolerable risk

Another way to arrive at tolerable risk targets is to examine accident statistics and see how various countries and types of industries normally perform.

If we can obtain a consistent method of measuring accident rates then we can see how we are doing in comparison to others. The following data from European studies shows how different countries measure up on an approximately consistent scale of measurement. The scale of measurement for injury rates is based on relating accident records to the number of workers employed in an industry.

C.2.1   Workplace injury in Europe and the USA, 1996

The following table shows the rates of fatal and over 3-day injury per 100 000 workers or employees:

Country	Rate of Fatal Injury	Rate of Over 3-Day Injury	Employed People Covered
EU average	3.6	4200	
Great Britain	1.9	1600	Workers
Sweden	2.1	1200	Workers
Netherlands	2.7	4300	Employees
USA	2.7	3000	Workers
Germany	3.5	5100	Workers
Italy	4.1	4200	Workers
Spain	5.9	6700	Workers
Portugal	9.6	6900	Employees

The above and following data is taken from an HSE report issued 27 September 2000 and is based on a Eurostat publication 'Accidents at Work in the EU in 1996 – Statistics in FOCUS, Theme 3–4/2000' HSE supplied data on USA and Netherlands from their own studies.

C.2.2   Industry sectors for the EU average and Great Britain, 1996

The following table shows the rates of fatal and over 3-day injury per 100 000 workers:

Industry Sector	EU Average		Great Britain	
	Fatal	Over 3 Day	Fatal	Over 3 Day
Construction	13.3	8000	5.6	2700
Agriculture	12.9	6800	10.8	2000
Transport, storage and communication	12.0	6000	1.2	2400
Electricity, gas and water	5.7	1600	1.4	1700
Manufacturing	3.9	4700	1.4	2200
Within manufacturing				
Food, beverage, tobacco	4.7	6600	0.9	Na
Non-metallic mineral products	8.1	6500	1.4	Na
Basic metals and metal products	7.7	8500	2.1	Na
Wood and wood products	8.5	10800	4.9	Na
Wholesale, retail trade and repairs	2.5	2400	0.4	1300
Financial, real estate, business	1.6	1600	0.3	700
Hotels and restaurants	1.1	3500	0.3	1500

Eurostat's results and a study by HSE show that the rate of fatal injury in Great Britain is one of the lowest in Europe, and is lower than that of the USA. Other charts in this series allow comparisons between individual countries.

Note that the data here identifies significant injuries causing over 3 days of absence from work. This data may be useful for evaluating tolerable rates for accidents of lower severity levels.

C.2.3 Qualitative methods of establishing safety

The EN 1050 standard describing the risk assessment procedure requires that the design team decide if the machine is safe. Safety of a machine is described by EN 292-1 paragraph 3.4 as the ability of a machine to perform all its functions 'without causing injury or damage to health'.

- EN 1050 then goes on to describe how it might be decided that the machine is safe by making comparisons with existing machines (see Section 3.7.5).

The standard notes that the assessment must still consider specific conditions of use and quotes an example of the possible differences in risk when using a band saw to cut meat compared with using it to cut wood.

This simple comparative approach may have its limitations but where well-established and proven machinery designs have been in existence for a long time it is logical to use them as a standard of safety.

Another qualitative approach is to use the simple risk matrix method where qualities are attributed to risks such as high, medium, low or negligible. We have already seen this in Chapter 3. Some industries employ risk priority numbering (RPN) scales where the risks are only acceptable if they fall below a certain reference level. Effectively the tolerable risk levels are decided at the time of generating or calibrating the scale. Examples can be found in the USA-based Robot Industries Association (RIA) risk matrix and in the version published in the USA Association of Manufacturing Technologies Report ANSI B11-TR3. This report discusses tolerable risk concepts to be used in risk assessment procedures.

Lastly it is worth noting the qualitative methods of SIL determination described in IEC 61508 part 5 where risk graphs are detailed. More specific details are to be found in IEC 61511 but these are mainly for use in the process industries. The graphical method of deciding what class of risk reduction is needed, if any, effectively decides what risks are acceptable in the same way that this has been done for the risk matrix. The same effect can be seen in the category selection chart in EN 954-1 as we have described in Chapter 5.

Appendix D

Notes on PUWER

This section supports material in Chapter 2 and relates to many of the protection features described in this manual. Our notes here are for indicating general points only. Readers should always refer directly to official publications for the actual text and scope of regulations.

D.1 Summary of some key points in The Provision and Use of Work Equipment Regulations 1998 (PUWER)

These regulations apply to the UK but are useful as a guide to good practices in the workplace in any country.

Part 1 consists of Regulations 1–3 dealing with scope:

- Defines a very wide scope of work equipment. Includes hand tools, single machines such as drills or photocopiers or even dumper trucks and combine harvesters.
- Lifting equipment such as hoists and work platforms.
- Installations such as series of machines. This will cover production and assembly lines as well as processing machines.
- Applies wherever work is done by the employed including self-employed persons. Includes offshore installations.

Part 2 consists of Regulations 4–24 dealing with general matters:

- Regulation 4: Suitability of equipment.

Requires employers to ensure that work equipment is suitable for the purpose for which it is intended. This deals with the safety of work equipment from the aspects of:

- Its initial integrity
- The place where it will be used
- The purpose for which it will be used.

These requirements link to the MHSWR regulations that require an employer to carry out a risk assessment on the workplace and work activities.

- *Regulation 5:* Maintenance

  - Requires efficient maintenance of machines to keep them in 'good repair'. Machines must not be allowed to deteriorate such that they put people at risk.
  - Frequency of maintenance to be arranged to allow for intensity of use and operating environment.
  - Maintenance to be managed on a systematic basis.

- *Regulation 6:* Inspection

  - Requires employers to carry out inspections of machines after installation or assembly at a new location.
  - Requires inspection of dangerous machines at regular intervals and after any exceptional circumstances.
  - Defines scope of inspections.
  - Requires inspection records with appropriate details.

- *Regulation 7:* Specific risks

  - Requires employers to ensure that only those given the task of using the machine be allowed to operate it.
  - Repairs, modifications, maintenance or servicing to be done only by competent persons assigned to the job.
  - Employers to ensure such persons are trained for the work.

- *Regulation 8:* Information and instructions

  - Adequate health and safety information to be provided by the employer relevant to the machines being used.
  - Written instructions to be provided for operation and maintenance. Duties of suppliers to provide instructions.
  - Conditions of use and conclusions based on experience to be made available.

- *Regulation 9:* Training

  - Employers are required to provide adequate training for employees on the use of the equipment.
  - Supervisors are to be trained on methods and risks in the work.

- *Regulation 10:* Conformity with EU Directives

  - Requires employers to ensure that equipment in use conforms with relevant EU Directives, e.g. Machinery Directive.
  - Applies to units supplied after 1 January 1993 and also applies to second-hand equipment supplied from outside the EU.
  - Effectively requires all employers to see that only CE marked equipment is being used.

- *Regulation 11:* Dangerous parts of machinery

  - Requires employers to ensure protection is provided against dangerous parts of any machine in the workplace.
  - Methods of protection begin with fixed guards where practicable and otherwise this requires that other types of guards or protection devices be fitted.

- Provide push sticks or jigs or similar protection devices where practicable, but where these measures are not enough also.
- Provide information for use, instruction, training and supervision.

- *Regulation 12:*   Protection against specified hazards

  - Requires employers either to prevent an employee being exposed to specific hazards or where this is not practicable to take measures to ensure that exposure is adequately controlled.
  - Does not accept that the use of personal protective clothing is sufficient but requires measures to minimize the effects of hazards and to reduce the likelihood of hazardous events. Effectively these are risk reduction measures as discussed in the manual.
  - Hazards are listed to include falling or ejected articles, rupture or disintegration of parts, fire or overheating, discharge of gas or vapors, unintended explosions.

- *Regulation 13:*   High or very low temperatures

  - Requires protection to prevent injuries due to very high or low temperatures used in the work equipment.

- *Regulation 14:*   Controls for starting or changing operating conditions

  - Machinery controls must ensure that it is not possible to perform any significant changes in conditions such as staring or speed changing except by deliberate action.

- *Regulation 15:*   Stop controls

  - Requires controls to bring the equipment to a safe condition in a safe manner.
  - Stop controls to have priority over any other controls.
  - Where necessary stop controls are to switch off all sources of energy after stopping the work equipment.

- *Regulation 16:*   Emergency-stop controls

  - E-stops to be provided unless the stop controls under Regulation 15 already provide adequate stopping safety.
  - E-stops to be positioned at workstations and any other appropriate points as determined by the risk assessment.
  - E-stops should not be considered to be an alternative to safe-guarding.

- *Regulation 17:*   Controls

  - All controls to be clearly visible with appropriate marking.
  - Controls to be located so that no person is exposed to risk when operating them.
  - Controls to be positioned so that the operator can see that no other person is in a dangerous area of the equipment. If this is not practicable, systems of work are to be provided to ensure the equipment cannot be started when a person is in the danger area.

    – May require audible or visible signals that equipment is about to start.

    – If a person can be put in danger from starting or stopping of work equipment there must be adequate measures to allow time and suitable means for them to avoid the risks.

- *Regulation 18:*   Control systems

    – Employers to ensure that as far as possible controls are safe and are chosen by making due allowance for the failures, faults and constraints to be expected in the planned circumstances of use.

    – A control system is not considered safe unless:

        • It does not create any increased risk to health and safety
        • That faults should lead to a fail-safe condition and
        • That its operation does not impede the functioning of stop controls and E-stops.

- *Regulation 19:*   Isolation from sources of energy

    – Requires suitable and identifiable means for ensuring the equipment can be made safe by isolating it from power sources. This applies to the hazards of the machine operation during maintenance, setting, cleaning, etc. Electrical safety requirements for isolation come under 'Electricity at Work Regulations'.

    – It also requires measures to ensure that the power isolation is secure against being switched on by accident or by others. This implies devices such as locking devices and multiple key lock systems.

- *Regulation 20:*   Stability

    – Requires that work equipment is stable and cannot cause harm by toppling or moving under use. Typically this will mean bolting down or clamping of fixed position machines. Temporary structures may need additional stabilizing devices.

- *Regulation 21:*   Lighting

    – Requires suitable and sufficient lighting to be provided for any task.

- *Regulation 22:*   Maintenance operations

    – Whilst Regulation 5 requires maintenance to be considered an essential part of safety management this regulation requires that it must be possible to carry out maintenance operations without exposing persons to risk.

    – Where risks are involved appropriate protection measures must be taken.

    – This regulation involves the employer in providing protection measures such as temporary guards, limited movement controls, hold-to-run controls and key lock systems.

- *Regulation 23:*   Markings

    – This is a general requirement for any information relevant to safe working to be clearly displayed. Examples include clear signs for start and stop controls, maximum load warnings or maximum speed warnings on grinding wheels and cutters. Warning signs for isolation of the machine before maintenance will be essential.

- *Regulation 24:*   Warnings
    - This requires the provision of warning signs and warning devices such as audible alarms and flashing lights where these are needed as part of the protection measures established for the machine (typically through the risk assessment process).
    - Specifically requires that warnings given by warning devices shall be 'unambiguous, easily perceived and easily understood'.

Part 3 consists of Regulations 25–30 dealing with mobile work equipment: This refers to any equipment that travels between different locations; it includes wheeled or skid mounted movable equipment and it may be self-propelled, towed or remotely controlled.

- No employees to be carried on mobile equipment unless it has features ensuring this can be done safely.
- Measures are required to prevent equipment rolling over onto people.
- Describes the provision of roll over protective structures (ROPS).
- Describes the provision of restraining systems where needed to protect persons against crushing if mobile equipment rolls over.
- Describes a series of measures for safety of self-propelled vehicles including prevention of unauthorized starting, provision of braking and stopping, etc. Items such as field of vision and lighting are covered here.
- Deals with remote-controlled self-propelled work equipment. In particular the regulations require that such machines stop automatically when outside the range of control.
- Deals with the risks of drive shafts delivering power from mobile machines to accessories. Specifically this would apply to 'power take-off shafts' in agricultural machinery. Requires guarding of shafts and measures to avoid harm due to seizure of shafts, which can lead to break up of machines or ejection of parts.

Part 3 consists of Regulations 31–35 dealing with power presses:

- Requires examination of any power press (except certain exemptions for smaller presses) before it is put into service. It also requires regular and thorough examinations to be done at least every 12 months for presses with fixed guards and at least every 6 months in other cases.
- Examinations to be done after any incident that could jeopardize the safety of the press.
- Requires persons making the examinations to report any defects uncovered by inspections to the employer.
- Reports on the results of examinations to kept available for inspection by authorities for 2 years. Certificates of inspection are to be placed at or near the power press.
- Guidance on the regulations is contained in the HSE publication: 'Safe use of power presses. PUWER 1998 as applied to power presses. Approved code of practice and guidance L112' (see Appendix A).

Part 4 consists of miscellaneous regulations dealing with the legal scope and does not impact on the technical material of the first three parts.

Appendix E

Guide to fault tree analysis

This section outlines the principles of fault tree analysis (FTA). The techniques of FTA are very helpful for providing a simple representation of the causes of a hazardous event. They are used in conjunction with failure rate estimates and probability estimates to arrive at predicted rates for the hazardous event.

FTA begins with a 'top event' that is usually the hazardous event we are concerned with, for example, 'explosion'. The 'tree' is then constructed by developing branches from the top-down using two basic operators: 'AND gate' and 'OR gate'. The logic gate symbols are shown in Figure E.1.

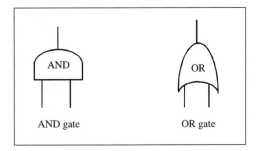

Figure E.1
Logic gate symbols for FTA

The logic gates allow the contributing causes of the top event to be set out and combined according to the simple rules of AND or OR (e.g. see Figure E.2).

E.1 Functions of the gates

AND gates are used to define a set of conditions or causes in which all the events in the set must be present for the gate event to occur. The set of events under an AND gate must meet the test of 'necessary' and 'sufficient'.

Here 'necessary' means each cause listed in a set is required for the event above it to occur; if a 'necessary' cause is omitted from a set, the event above will not occur. 'Sufficient' means the event above will occur if the set of causes is present; no other causes or conditions are needed.

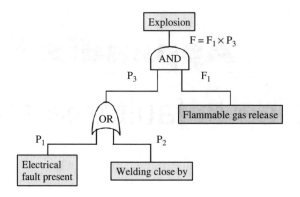

Figure E.2
Example of basic fault tree

OR gates define a set of events in which any one of the events in the set, by itself, can cause the gate event. The set of events under an OR gate must meet the tests of 'sufficient'.

The information about each event is described as either: P = Probability of the event occurring, or f = Frequency of the event, or f × t = Duration of the event. From which the following combinational rules are obtained:

Inputs	Gate	Operation	Output of the Gate
P_1, P_2	AND	$P_1 \times P_2$	P
P_1, f_1		$P_1 \times f_1$	f
$(f_1 \times t_1), (f_2 \times t_2)$		$(f_1 \times f_2)(t_1 + t_2)$	f
P_1, P_2	OR	$P_1 + P_2$	P
f_1, f_2		$f_1 + f_2$	f

The combinational rules allow the information known about each individual event to be combined to predict the frequency of the top event and the intermediate events.

E.2   Event symbols

Event symbols used in FTA are as shown in Figure E.3.

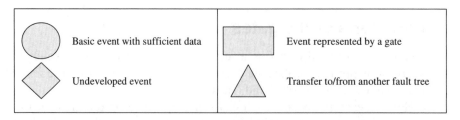

Figure E.3
Event symbols

These provide a means of classifying events. Basic event is the limit to which the failure logic can be resolved. A basic event must have sufficient definition for determination of an appropriate failure rate. Undeveloped events are events that could be broken down into subcomponents, but, for the purposes of the model under development, is not broken down further. An example of an undeveloped event may be the failure of the instrument air supply. An undeveloped event symbol and a single failure rate can be used to model the instrument air supply rather than model all the components. FTA treats undeveloped events in the same way as basic events. Rectangles are used above gates to declare the event represented by the gate. Transfer gates are used to relate multiple fault trees. The right or left transfer gates associate the results of the fault tree with a "transfer in" gate on another fault tree. House events are events that are guaranteed to occur or guaranteed not to occur. House events are typically used when modeling sequential events or when operator action or inaction results in a failed state.

The fault tree construction proceeds by determining the failures that lead to the primary event failures. The construction of the fault tree continues until all the basic events that influence the top event are evaluated. Ideally, all logic branches in the fault tree are developed to the point that they terminate in basic events.

Here's an example of a fault tree used to troubleshoot a problem when a car fails to start (see Figure E.4).

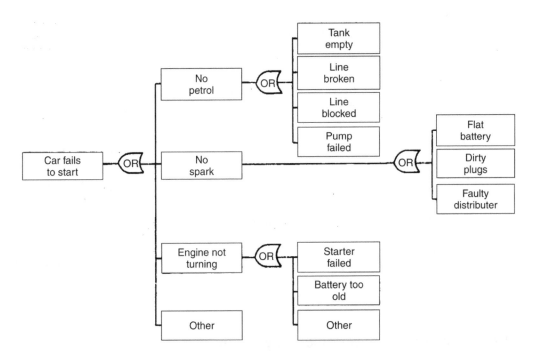

Figure E.4
Fault tree used for breaking down a problem into possible causes

E.3   Adding risk reduction measures in FTA

So far we have seen how useful fault tree analysis can be for anglicizing the risk of a known top event. The next step is to build in the possible risk reduction measures and predict the new risk frequency for the top event. It's easy to do this using the general approaches shown in Figures E.5 and E.6.

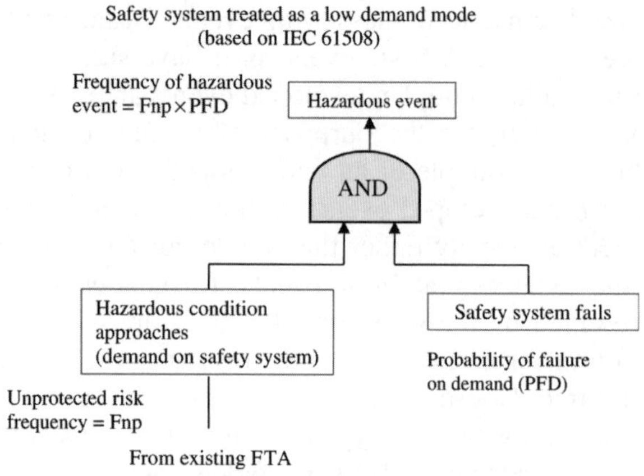

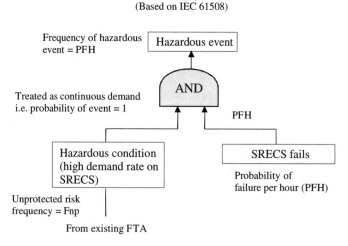

Figure E.5
Generic method for adding SREC protection into a fault tree model when operating in the low demand mode

Figure E.6
Generic method for adding SRECS protection into a fault tree model when operating in the high demand mode

E.3.1   Method 1: Low demand systems

Method 1 is used for 'low demand rate' safety systems where the demand on the system to be available to respond to a hazardous event is at a low rate, typically once per year (as defined by IEC 61508). Alternatively low demand can be defined as a system where there are no more than two demands on the average between the scheduled proof tests (as defined by IEC 61511).

In Figure E.5 the protection in the form of SRECS has an expected probability of dangerous failure per hour (PDF). This probability feeds into an AND gate with the predicted frequency of the hazardous event (demand rate F_{np} per hour) assuming no protection. The additional AND gate reduces the frequency of the hazard to $F_{np} \times PDF$.

E.3.2 Method 2: High demand or continuous mode systems

Where the demand rate is high or continuous the hazardous event rate is based solely on the failure rate of the safety system, expressed as the probability of failure per hour (PFH). This method assumes that the probability of demand on the safety system can be approximated to 1. Hence the hazardous event rate is defined by the probability of failure per hour of the safety system.

Practical exercises

Exercise 1: Calculating risk parameters

This practical exercise supports Chapter 1.

Subject	Calculation of individual risk (IR). Interpretation of relative values for a decision on risk reduction
Objective	To assist participants to become familiar with risk criteria and their interpretation
Relevance	An important component of the quantitative methods for evaluation of risks. It is helpful to have the ability to estimate a given risk in terms that can be compared with industrial accident statistics
Starting information	A hazard study of an automatic assembly machine has found that there is a risk that an operator could become snagged in the moving parts with the possibility of a permanent injury to hand or arm. The plant has five such units operating under the control of a shift team manned by an average of three persons per shift. The problem was referred for hazard analysis, which found that if no guards are fitted the likelihood of an accident is once per year per machine
Task detail	**Practical is for individual participants** *Task 1:* Calculate the individual risk per worker in terms of the probability of serious accident per year. Compare this with a suggested tolerable risk target of 1×10^{-5} for an average work environment. Decide the quantitative risk reduction factor that would be needed from a protection system to be fitted to the machine *Task 2:* Assume the possible injury from the machine is limited to being an accident that causes more than 3 days absence from work. Calculate how the risks to the combined workforce compare with a typical average for the UK of 1600 3+ day injuries per 100 000 workers employed
Time allowed	Thirty minutes, including time for discussion of answers

Exercise 2: Developing safety requirements

This practical exercise uses subjects covered in Chapters 2, 3 and 4.

Subjects	Selection of a protection measure and defining functional requirements. Decision on safety category
Objective	To assist participants to become familiar with developing functional requirements and safety integrity requirements
Relevance	It is helpful to have the ability to estimate a given risk in terms that can be compared with industrial accident statistics
Starting information	A hazard study of a large steel strip production line has established three points in the line where a worker could become entrapped in the moving parts with the possibility of a permanent injury or death. There is a need to access these points periodically to clean out debris. This would only occur when the line has been stopped and will typically occur once per day. The task requires an operator to have full bodily access to the production line
Task detail	*Step 1:* Consider and write down the options that might be used to protect the operator from injury and decide which method you would recommend *Step 2:* Write down a clear description of the functional requirements of this protection measure. Include a sketch if it helps *Step 3:* Decide the required safety category for the safety-related electrical parts of the protection system. Write down the reasons for this decision in terms of the EN 954 guide chart shown in Chapter 5
Time allowed	Thirty minutes, including time for discussion of answers

Exercise 3: Risk assessment and risk reduction

Subject	Risk assessment exercise on a garden edge trimming tool
Objective	To gain basic experience with the EN 1050 risk assessment and EN 292 risk reduction procedures
Relevance	All machinery safety projects should be subjected to this type of procedure. Provides a simple practice routine. This example is typical but many variations can be devised
Starting information	Suppose you are the first person to have invented a portable electrical hand tool for trimming the edges of lawns and paths in parks and gardens. It has a powered spinning hub on the end of a long handle. The hub carries small blades or has lengths of nylon cord You are required to do a preliminary risk assessment combined with a risk reduction exercise following the steps described in EN 1050 and including the risk reduction procedures described by EN 292-1 and as summarized by Figure 3.9
Task detail: Practical is best done by two persons working as team	*Task 1:* Set up a risk assessment record sheet and begin by determining the limits of the machine and its application environment (see Section 3.3). Then identify as many hazards as possible and itemize them on your record sheet *Task 2:* Using the risk matrix example given below (as per Figure 3.8) mark in the location of each risk assuming there are no safety measures *Task 3:* Now apply the risk reduction steps shown in Figure 3.9 to each hazard in turn and record your suggested measures. These will include the obvious design measures such as simple guard over the blades and hold-to-run controls *Task 4:* Update the risk matrix by marking in the new position of each hazard item number
Time allowed	Thirty minutes, including time for discussion of answers

Times per Year	Frequency per Year	Consequence			
		Minor: 1	Significant: 2	Major: 6	Catastrophic: 10
1–10	**Frequent: 10**				
0.1–1	**Probable: 8**				
0.01–0.1	**Possible: 4**				
0.001–0.01	**Unlikely: 2**				
0.0001–0.001	**Remote: 1**				

Table for recordings the risk assessment:

Haz. No.	Hazard or Event	Possible Causes	Consequences	Initial Risk Value	Design Features or Safeguards to Reduce Risk			Final Risk Value. Acceptable?
					By Design	By Safeguard	By Information	
1								
2								
3								
4								
5								
6								
7								
8								

Exercise 4: Application of fault tree analysis

See Appendix E for a primer on this subject before attempting this exercise.

Subject	Application of fault tree analysis (FTA) to predict an accident rate before and after application of safety system
Objective	To gain basic experience with using fault tree analysis to describe and evaluate proposed safety measures
Relevance	Machinery safety projects frequently require the basis of a safety measure to be described in logical and quantitative terms. Fault tree analysis is a useful method for graphically describing problems
Starting information	A simplified grinding machine example is used. The machine has to remove rough surface material from a workpiece clamped to a moving table. The machine is initially considered without any cover over the grinding wheel or work area (see Figure Ex. 1) An operator is required to place and remove a steel object or workpiece for surface grinding and to clamp the item to the sliding table. The operator then starts the machine and the grinding wheel runs up to speed. The operator then switches the table into automatic traversing with a fixed height and advance rate between the table and the wheel. Grinding strokes take place until the task is completed, the operator switches off the grinder and removes the workpiece. This task is repeated at frequent intervals There is a risk of injury to the operator from flying particles causing eye damage and also there is a risk that the workpiece will fly out of the machine if it is not clamped properly and if it is oversized. An oversized piece may cause the wheel to break up and the operator could be struck by flying parts. The operator has been provided with safety glasses and instructed to wear them at all times. Hence the analysis model must allow for a degree of probability that the operator will forget to put on the safety glasses Once the basic fault tree has been drawn, the next step will be to examine the effect of providing an interlocked movable guard as per Figure Ex. 2
Task detail: Practical is for individuals. However it helps to compare details with others	*Task 1:* Draw a fault tree to show a top event as an eye injury to the operator. The logic for an eye injury is the combination of minor injury due to projectiles with the probability that the operator will not be wearing goggles. Show these as contributors to the top event. Expand the tree downwards to show the logical causes of the minor injury event and their relationships to it *Task 2:* Apply the fault rates as shown in the table below to your FTA diagram to compute the overall eye injury rate predicted for this scenario. Indicate which items can be estimated as probabilities and which items are frequencies *Task 3:* Now modify your fault tree to show the presence of the interlocked guard. It can be drawn in as a complete "safety system" in one block. There is no requirement in this exercise to detail the parts of the safety system. The method of applying the guard safety system will be based on the high demand mode principle shown in Appendix E. Also assume that the operator will no longer be required to wear goggles *Task 4:* Calculate the new injury rate for the operator. We will use a figure of 10^{-5} dangerous failures per hour
Time allowed	Forty minutes including time for discussion of answers

Table of data

No. of pieces per hour	10
Probability that operator is in danger zone	0.2
Chance of wheel breaking if oversize piece is fed	0.02
Probability of defeat or failure per hour of guard system	10^{-5}
Probability of oversized piece	0.01
Probability of loose clamp	0.002
Chance of projectiles hitting area of eyes	0.3

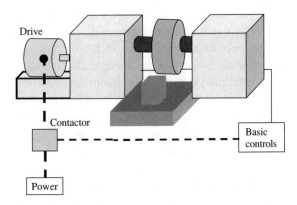

Figure Ex. 1
Outline of grinding machine with table and workpiece

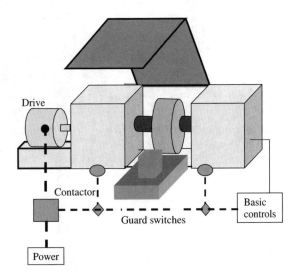

Figure Ex. 2
Grinding machine with interlocked safety guard added

Exercise 5: Outline design of a safety system

Subject	Draw up a block diagram representation of a safety system for a particular application. Show the critical features needed to satisfy category 3
Objective	This exercise basic understanding of safety system features
Relevance	Safety categories in Chapter 5
Starting information	Figure Ex. 3 is a very approximate representation of an automated cutting table fitted with a pneumatically driven feeding system for components. An operator loads a batch of blank steel sheets onto an automatic feeding table. An elevator and transfer mechanism feeds each plate into the machine and a stacker machine places the finished plates on the exit stack Once loaded, the machine is started from a panel located outside the protected area. A light curtain provides area protection by using an emitter/receiver array with reflectors at each corner to fully enclose the machine. If a person attempts to enter the area the machine will stop and it has to be restarted at the control panel. For extra protection, an E-stop button is located on the control panel The machine is operated by a single electrical drive whilst solenoid valves operate the handling systems. All sequencing and control is handled by a PLC. All motion can be safely stopped by removing power to the solenoids and the motor
Task detail: Practical is for individuals	*Task 1:* Draw up a simple block diagram representation of the safety system comprising a light barrier sensor and control unit, E-stop, logic control and final element controls. The design is required to be compliant with Category 3 principles, as laid down by EN 954-1. Indicate on your design diagram the arrangement and features that should be included to satisfy this requirement. If needed add notes to justify or explain the features you would suggest
Time allowed	Thirty minutes, including time for discussion of answers

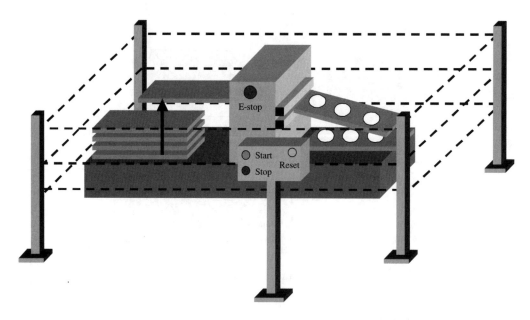

Figure Ex.3
Automated cutting table with area protection

Exercise 6: Calculating safety distance for a light curtain application

Subject	Calculation of safety distance as function of machine parameters and AOPD design
Objective	To help clarify the application of safety distance calculation methods
Relevance	Supports Chapter 7. Fundamental requirement for all electrosensitive protection elements, in particular light curtains
Starting information	Assume that a perpendicular light curtain is to be used for point of operation guarding of a bottle-filling station on moving production line (see Figure Ex. 4). The machine runs automatically but requires an attendant who will be needed to remove a broken or overturned bottle from time to time. The following data applies: Response time of the light curtain: 50 ms; machine stopping time: 200 ms; minimum object sensitivity of the light curtain: 35 mm
Task detail: Practical is for individuals	*Task 1:* Calculate the minimum safety distance allowing for a safety margin of 30% on the machine stopping time. Then check whether or not this distance will allow a person to be undetected between the light curtain and the machine. If so, decide what measures should be taken to ensure safety *Task 2:* Suppose that a horizontal light curtain is to be considered as an option. Calculate the safety distance for the outermost light beam if the horizontal curtain is mounted at 500 mm above ground
Time allowed	Thirty minutes, including time for discussion of answers.

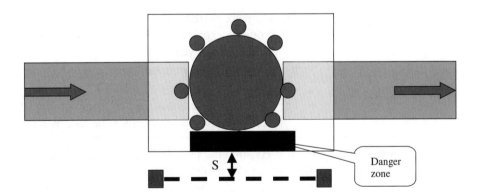

Figure Ex.4
Representation of a bottling line with filling machine. Danger zone and proposed light curtain are shown

Answers to practical exercises

Exercise 1

Answers to task 1	*Task 1:* Calculate the individual risk per worker in terms of the probability of serious accident per year. Compare this with a suggested tolerable risk target of 1×10^{-5} for an average work environment. Decide the quantitative risk reduction factor that would be needed from a protection system to be fitted to the machine. There are six workers sharing five potential accidents per year. So for one worker the individual risk per year is 5/9 accidents per year = 0.55 per year If the target for an acceptable risk is 1×10^{-4} the machines are too dangerous by a factor of $0.55/10^{-4} = 550$ times too dangerous For the machines to satisfy an IR target of 1×10^{-4} the protection systems would have to reduce the risk by a factor of 550
Answers to task 2	*Task 2:* Assume the possible injury from the machine is an accident that causes more than 3 days absence from work. Calculate how the risks to the shift workforce compare with a recorded average for the UK of 1600 injuries per 100 000 workers employed. There are nine persons employed. The number of accidents expected without protection on the machines is five per year This would convert to $5 \times 100\,000/9 = 55\,555$ per 100 000 workers This is higher than the average by a factor of $55\,555/1600 = 350$ times higher For the machines to present a risk no higher than the average injury rate for the industry, the protection systems would have to reduce the risk by a factor of 350

Notes	In Task 1 the target IR of 10^{-4} is an approximate figure based on being 10 times higher than the Alarp boundary fatal injury rate that has been suggested by HSE for planning of industrial sites. It has no official endorsement. The calculated IR indicates the scale of improvement in safety that would be needed for the machines to meet a target that might be justified as being as low as reasonably practicable
	In Task 2 the predicted average accident rate (scaled to 100 000 workers) for the plant will be calculated from the total number of persons employed in the plant and total of probabilities of accidents per year. These figures (plant average and national averages) may assist the design team to establish what is a reasonable and tolerable risk rate for the machines in question
	The answers obtained by either method indicate that a safety improvement factor of typically 300–500 will be needed for the machines. This provides a basis for the design of guarding systems. In Chapter 9 of the workshop we shall see how this quantitative factor can define the SIL rating of the safety-related controls

Exercise 2

Information	A hazard study of a large steel strip production line has established three separate points in the line where there is a need to reach into the machinery periodically to clean out debris. The cleaning can only be done when the line has been stopped and will typically occur four times per day The whole of the production line is enclosed by a 2 m high steel fence placed at 1 m distance from the edges of the machine. The fence has locked gates used only for maintenance access under a controlled key interchange procedure If the machine is moving or starts whilst the cleaning task is being done the worker could become trapped in the moving parts with the possibility of a permanent injury or death
Answer for Step 1	*Step 1:* Consider and write down the options that might be used to protect the operator from injury and decide which method you would recommend *Option 1:* Access the parts using the maintenance gate and walk along the affected areas. Not acceptable because the procedure is slow, probably means climbing over parts and needs maintenance staff assistance. Also presents a risk that a person can be in the danger area when the machine is started *Option 2:* Modify the fence to place it closer to the points where cleaning is needed. Provide a sliding panel with a guard-monitoring switch. This allows the operator to reach into the machine. Machine will stop when the sliding panel opens. Not acceptable because there is a danger that the machine will still be running down when the operator reaches into it. Undesirable also because this will cause unscheduled emergency stopping of the mill *Option 3:* As Option 2, but provide an interlocked guard ('stop locked') with run-down timer or zero-motion detector. The guard cannot be opened until the machine has stopped. The machine cannot be restarted until the guard has been closed. Acceptable because entry is only possible under safe conditions. Lost time is minimized and unscheduled stops are avoided *Option 4:* As Option 2, but leave a permanently clear access point and fit a light curtain across the space. Machine will stop if operator attempts to reach into the space before it has stopped. Not acceptable because safety distance needed to allow the machine to stop will be too great to allow the job to be done. Leads to unscheduled E-stops and is likely to be more expensive than Option 3
Answer for Step 2	*Step 2:* Write down a clear description of the functional requirements of this protection measure. Include a sketch if it helps Each access point is to have a sliding panel that cannot be physically removed without tools The panel is to be placed opposite the cleaning point but shall be designed to prevent persons climbing into the machine via the opening it provides The sliding panel is to be held shut by a solenoid-operated bolt lock that must be energized to release the panel. A pushbutton is to be provided for the operator to operate the bolt when required, and a lamp indicator is required to show a green light when the release condition is available It shall not be possible for the solenoid to be energized unless the power to the drives of the machine has been switched off and there is no movement

	of the critical rotating parts in the area of access. A zero-motion detector is to be fitted to the rotating element in the machine. Where these elements are different at each cleaning point a separate zero-motion sensor should be used at each point The sliding panel shall be fitted with a limit switch to detect when it is closed. The switch contacts shall be closed only when the door is closed and the actuator shall be a coded tongue. The power to the machine drives shall be tripped whenever the limit switch contacts are open
Answer for Step 3	*Step 3:* Decide the required safety category for the safety-related electrical parts of the protection system. Write down the reasons for this decision in terms of the EN 954 guide chart shown in Chapter 5. The protection system should be rated as Category 3 The reasons are given in terms of the risk parameters shown in the EN 954 chart: Severity of injury: S2 Frequency and/or exposure: F2. It is possible to consider the frequency/exposure as F1 but due to the fact that there are three locations on the machine being accessed regularly the F2 option is probably the best decision Possibility of avoidance: P1; the hazard will always be clear to the operator
Conclusions	This exercise has concluded that an interlocked guard should provide protection with a zero-motion detector. The safety-related parts should be Category 3. There is some scope for evaluation of whether or not a fault-tolerant bolt lock system is needed since it is backed up by a pushbutton and a guard switch

Exercise 3

Subject	Risk assessment exercise on a garden edge trimming tool
Task detail: Practical is best done by two persons working as team	*Task 1*: Set up a risk assessment record sheet and begin by determining the limits of the machine and its application environment (see Section 3.3). Then identify as many hazards as possible and itemize them on your record sheet
	The limits of the machine include the complete edge trimming tool, any special key or wrench needed to fit the blades or cord, the power cable and the instruction set and a set of goggles. The limits of use refer to its use for trimming lawn edges and surfaces, the trimming or low weeds and shrubs and the possible misuse to raise it to head height or above to cut foliage from trees and hedges
	Hazards are recorded as numbers 1–8 on the risk assessment form supplied
	Task 2: Using the risk matrix example given below (as per Figure 3.8) mark in the location of each risk assuming there are no safety measures
	Each risk has been marked on the risk assessment form before any measures have been taken into account. Numbered circles for each risk are placed on the risk matrix as shown below
	Task 3: Now apply the risk reduction steps shown in Figure 3.9 to each hazard in turn and record your suggested measures. These will include the obvious design measures such as simple guard over the blades and hold-to-run controls. See the table for answers
	Task 4: Update the risk matrix by marking in the new position of each hazard item number. See the risk matrix with new risk values as numbered circles at the end of the arrows
Time allowed	Thirty minutes, including time for discussion of answers

Times per Year	Frequency per Year	Consequence			
		Minor: 1	**Significant: 2**	**Major: 6**	**Catastrophic: 10**
1–10	**Frequent: 10**	5	1	8	
0.1–1	**Probable: 8**	1	2 4	3	
0.01–0.1	**Possible: 4**	5	2	7 6	
0.001–0.01	**Unlikely: 2**		3	4 8	
0.0001–0.001	**Remote: 1**			7 6	

Records of the risk assessment

Haz. No.	Hazard or Event	Possible Causes	Consequences	Initial Risk Value	Design Features or Safeguards to Reduce Risk			Final Risk Value. Acceptable?
					By Design	**By Safeguard**	**By Information**	
1	Spinning cutter makes contact with foot or leg	Tripping or inattention Dropping the tool Working steep banks	Serious cuts to foot or leg	20	Avoid using steel blades. Use nylon cord. Reduces consequences	.	Advise use of strong footwear (PPE)	8 Yes
2	Flying stones	Loose stones on ground	Minor injuries to public, damage to windows or vehicles	16	Avoid using steel blades		Keep people away. Use warning signboard in public areas	8 Yes
3	Stones and wood chips hit operator in the face	Working loose surfaces or undergrowth Misuse on trees and bushes	Eye damage	48	Long handle to keep working parts away from face	Fit a guard to shield most of the cutting edge	Warning notices on machine. Provide goggles with product	4 Yes
4	Machine thrashes around on the ground	Operator drops running machine and motor keeps running Operator trips over with machine in hand	Serious cuts and bruises	16	Control button in the handle to be 'hold-to-run'. Handle design with grips but with easy release	Start/run control to be positive action with spring release. Design to avoid jamming or tampering		4 Yes

Haz. No.	Hazard or Event	Possible Causes	Consequences	Initial Risk Value	Design Features or Safeguards to Reduce Risk			Final Risk Value. Acceptable?
					By Design	By Safeguard	By Information	
5	Pulling or wrenching	Cutter hub tangles with vegetation	Arm or back strain	10	Hub to be smooth profile. No parts to snag. Low inertia hub designed to stall			4 Yes
6	Electric shock	Operator error leads to cutting of electrical cord. Exposes wires	Possible electrocution	24	Cable entry at handle well away from cutter	Always use earth leakage protection on power supply	Warning notices in the instructions	6 Yes
7	Electric shock	Using machine in wet conditions	Possible electrocution	60	Double insulation of motor. Waterproof start run control and sealed electrical connections	Always use earth leakage protection on power supply	Warning notices in the instructions	6 Yes
8	Drawing in or tangling of clothes	Exposed drive shaft or hub	Drags machine against the body, could cause significant harm	20		Enclose all drive shafts and moving parts. Guards may be removable		4 Yes

Exercise 4

Subject	Application of fault tree analysis (FTA) to predict an accident rate before and after application of safety system.

Task 1: Fault tree logic only.

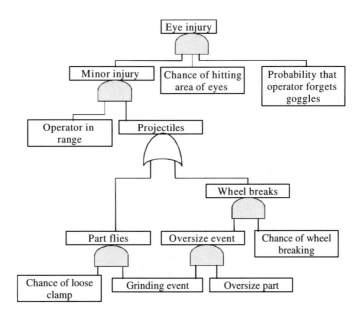

Task 2: Calculation of eye injury rate by adding data and calculating rates at each logical point. The results indicate that an eye injury is expected to occur at the rate of approximately twice per year if the plant operates 24 h per day. Alternatively this is one eye injury every 1.5 years per operator.

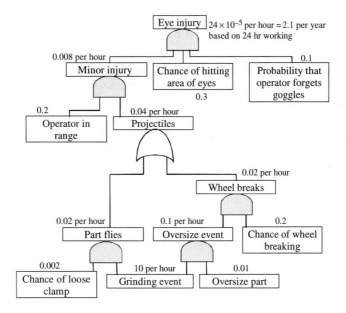

Task 3: Add the protection function of an interlocked guard. In this case the calculation has been treated as a high demand or 'continuous' mode of operation as defined by IEC 61508. This method assumes that the probability of demand on the safety system can be approximated to 1. Hence the accident rate is defined by the probability of failure per hour of the safety system. In this example we have stated that this figure will be 10^{-5} failures per hour. This is a conservative figure and represents a minimum performance level for a safety system.

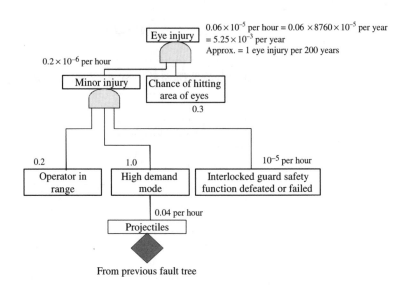

From previous fault tree

Task 4: Calculate the new accident rate. The fault tree shows that the new rate will be approximately 1 eye injury per 200 years. This is based on the assumption that the operator will not be expected to wear goggles if the interlocked guarding system is installed.

To further reduce this predicted injury rate we could consider improving the reliability of the safety system through frequent testing or more redundancy. Or we could reinstate the requirement for goggles but this might be an unrealistic approach given the presence of a complete guarding system.

Note: The alternative to using the high demand mode model shown above is to model the safety system failure as an event with an average duration of half the interval between proving tests of the safety system. Then the frequency and time period can be multiplied by the frequency of the 'projectiles event' using the formula given in Appendix E. This is the basis of the low demand mode model shown in Appendix E. Depending on the test interval used this will give a result that yields a lower accident rate than the high demand mode but the validity of the model would have to be carefully checked from first principles. For a detailed description of failure rate modeling, the reader should refer to IEC 61508, Part 6, Appendix B.

Exercise 5

Subject	Draw up a block diagram representation of a safety system for a particular application. Show the critical features needed to satisfy Category 3.
Task detail: Practical is for individuals	Draw up a simple block diagram representation of the safety system comprising a light barrier sensor and control unit, E-stop, logic control and final element controls. The design is required to be compliant with Category 3 principles as laid down by EN 954-1. Indicate on your design diagram the arrangement and features that should be included to satisfy this requirement. If needed add notes to justify or explain the features you would suggest.
Answer notes	The suggested arrangement for a Category 3 SRECS is shown overleaf. The following points may be relevant: The solution is shown in approximate block diagram form and therefore does not show all the connection details that would apply to a detailed circuit design. The proposed solution is a version based on hard-wired equipment. An alternative scheme using a programmable safety system with or without network-connected sensors and final elements could be proposed. For Category 3 each subsystem must be fault-tolerant. In the case of the light curtain this may be achievable by single-channel equipment using a high level of self-checking (diagnostic coverage). In the diagram we have shown a dual-redundant circuits in the control unit both carrying out frequent cyclic checks on the light beams. Hence any fault in a single beam will be detected and the safety system will trip, whilst a fault in a control circuit will not cause failure of the safety function due to the redundant channel. The E-stop is connected in series with the light curtain trip contacts from both channels. The E-stop switch has positive acting contacts and has dual-channel switches. A Category 3 approved monitoring relay module has been proposed where two independent input channels are arranged with internal cross-monitoring. The relays K1 and K2 cannot be latched in by the reset function unless all output devices are confirmed to be in the off position and both inputs circuits are closed. Once K1 and K2 are latched in the PLC outputs are able to operate the final elements under normal logic control. A feedback signal from the monitoring relay to the PLC allows the PLC to respond correctly to the status of the safety trip. The final elements that control all motion of the machine include pneumatic cylinders. Therefore the safety trip also applies to the master solenoids controlling air to a solenoid set performing sequencing under control of the PLC. The master solenoids are duplicated to ensure fault tolerance to Category 3 and position feedback has been included to confirm that the solenoids have de-energized before the safety function can be reset. Alternative designs for tripping out the pneumatic functions may be acceptable but at all times the possibility of failed final element devices must be taken into account.

Suggested arrangement for Category 3 protection based on hard-wired equipment.

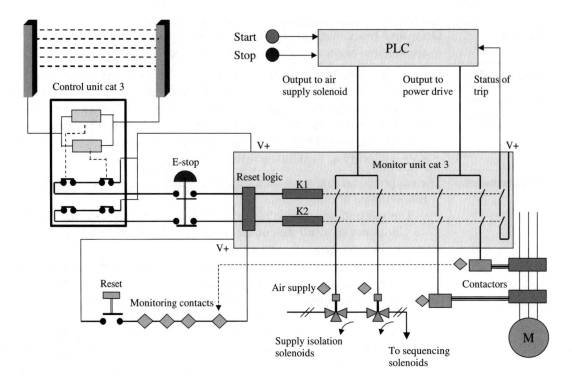

Exercise 6

Task 1 Assume that a perpendicular light curtain is to be used for point of operation guarding of a bottle-filling station on moving production line. The machine runs automatically but requires an attendant who will be needed to remove a broken or overturned bottle from time to time. The following data applies: response time of the light curtain: 50 ms; machine stopping time: 200 ms; minimum object sensitivity of the light curtain: 35 mm.

Calculate the minimum safety distance allowing for a safety margin of 30% on the machine stopping time. Then check whether or not this distance will allow a person to be undetected between the light curtain and the machine. If so decide what measures should be taken to ensure safety.

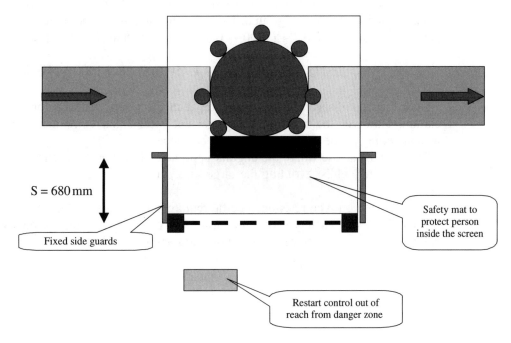

Figure An.1
Safety distance and additional safety measures

For a perpendicular light curtain where $d < 40$ mm the safety distance S is shown in Figure An.1 and is given by:

$$S = 2000\ T + (8 \times (d - 14))$$

unless S becomes greater than 500 mm. In this case T is the sum of the response time of the light curtain and the stopping time of the machine +30%.

$$T = 0.05 + (0.2 \times 1.3) = 0.31\ \text{s}$$

$S = 2000 \times 0.31 + 8 \times (35 - 14) = 620$ mm $+ 168$ mm $= 788$ mm. Since this figure is greater than 500 mm the alternative formula given in Chapter 7 can be used. Then

$$S = 1600 \times 0.31 + 8\ (35 - 14) = 496\ \text{mm} + 168\ \text{mm} = 664\ \text{mm}$$

This distance is large enough for a person to get between the light curtain and the danger zone. Hence additional measures are needed to protect against this possibility. Suggestions are:

1. Build fixed side guards to prevent side entry behind the light curtain.
2. Install a safety pressure mat to prevent machine running if a person stands in the danger zone.
3. Alternatively the pressure mat function could be performed by a horizontal light curtain across the area inside the vertical light curtain.
4. If it is acceptable for the machine to be restarted by hand the restart control, which also resets the machine trip, must be located where there is full view of the danger zone but where it cannot be reached by a person in the danger zone.

Task 2   Suppose that a horizontal light curtain is to be considered as an option. Calculate the safety distance for the outermost light beam if the horizontal curtain is mounted at 500 mm above ground. Using the formula for a horizontal light curtain given in Chapter 7 as $S = 1600\,T + (1200 - (0.4\,H))$ we note that $H = 500$. Then

$$S = 1600 \times 0.31 + (1200 - (0.4 - 500)) = 496 + 1000 = 1496 \text{ mm}$$

Note that this distance is much greater than the safety distance for the vertical light curtain. However the horizontal arrangement means that extra measures to protect a person getting inside the curtain may not be needed.

This may be a case where the angular approach will cover the need for the inside space to be protected whilst avoiding the large safety distance required for the horizontal light curtain.

The decision on type rests on specific requirements for each plant. Considerations will include the cost of each type of light curtain array balanced against productivity effects of creating a larger distance between the operator and the machine.

Index

Active opto-electronic protective device (AOPD), 158, 180

Active opto-electronic protective device responsive to diffuse reflection (AOPDDR), 158

As Low As Reasonably Practicable (ALARP) Principle, 12–13, 248–9

Control devices, 7–8, 134, 163
  basic functions, 101
  design safety, 96
  electrical interlocking, 143, 149, 150
  E-stop, 165–6
  failure modes, 99
  hazards of pushbuttons, 164
  hold to run, 132
  IFM Electronic range, 129–30
  mechanical trapped key interlocking, 152
  remote, 173
  risk reduction, 95
  Sipha 6 unit, 146
  two-hand, 130, 163, 173–5

Design techniques, 92–3
  considerations, 97
    characteristics of safety functions, 100–1
    common mode failures, 99
    failure modes of safety controls, 99
    general strategy, 98–9
    key points for basic control functions, 101–4
    safety objectives, 97
  EN 954-1, 93–4
    decide measures for risk reduction by control means, 95
    design safety controls, 96

hazard analysis/risk assessment, 95–7
  outcome, 97
  procedures, 94–5
  scope, 94
  specify safety control requirements, 96
  useful annexes, 94
  validation, 96
objective, 97
position on programmable systems, 93
safety categories, 104–5
  block diagram models, 106–8
  guidance for selecting, 108–10
  summary of each category, 104
  table, 105–6

Edge-sensitive devices, 157–8, 177
  benefits, 177
  disadvantages, 177
Electrical control interlocking devices, 143, 149, 150
Electronic E-stop monitors, 127–8
Emergency stop (E-stop) function, 102, 112–14, 165–6, 171
  see under stop functions
Enabling switch, 136
Engineering tasks:
  change control, 21
  safety lifecycle, 19
    conduct hazard identification exercise, 20
    design verification, 20
    detailed design/building, 21
    obtain information, 20
    outline design/identify subsystems involved in SRECS, 20
    provide design history file/technical file, 21

Engineering tasks: *(Continued)*
    risk reduction decisions, 20
    specify equipment/safety categories, 20
    use/maintain systems, 21
    validation, 21
Equipment choices, 17

Guard-locking systems/devices, 148, 171–2
    electrical control, 149, 150
    mechanical trapped key, 149, 152
    plus limit switches, 149
    safety issues, 150–1
    selection factors, 172–3
Guards/barriers, 138
    adjustable, 140
    automatic push-away, 140
    design requirements, 138
    distance, 140
    fixed, 138
        advantage, 138
        disadvantage, 138
        EN standards, 139–40
        features, 138
        typical applications, 143
    interlocking, 140–2
        advantages, 143
        disadvantages, 143
        electrical control, 143
        features, 142–3
        typical applications, 150
    movable, 140–2
    safe by position, 140
    sensing devices, 144–5
        guard-locking, 149–50
        mechanical limit switches, 144–5
        motion detection, 151
        non-contact (non-mechanical) actuation, 146

Hazard:
    analysis, 95
    combinations, 66
    electrical, 64
    identification, 62–3
    materials/substances, 65
    mechanical, 63–4
    neglecting ergonomic principles, 65
    noise, 65
    radiation, 65
    thermal, 64
    vibration, 65
Hazard study methods, 66
    deductive, 66
    Hazops, 69–70

    inductive, 67
    machinery concept hazard analysis, 67–8
    outcomes, 71
    results of activity analysis, 68
        error types, 69
Hazop studies, 69–70
    when to use, 70–1
Hold to run control device, 136

Laser scan sensors, 163
Light curtains, 160–2
    access guarding/perimeter guarding, 178
    application guidance notes, 180
        summary of techniques, 180
    area guarding, 177–9
    area guarding by scanning, 178–9
    benefits, 179
    choosing, 179–80
    design steps for AOPD:
        access control, 181
        angled approach, 185
        blanking, 187
        calculation of safety distance, 182–6
        define safety category, 181–2
        define safety function, 181
        define zone to be guarded, 181
        determine types of approach, 182
        horizontal approach, 185
        monitoring/stopping distances/speeds, 186–7
        muting of protective devices, 188
        perimeter/area guarding, 181
        perpendicular approach, 184–5
        point of operation, 181
        PSDI or single/double-break initiation, 187
        safety distance verification, 186–7
    point of operation guarding, 181
    terminology:
        AOPD, 180
        ESPE, 180
        minimum object sensitivity, 180
        muting/blanking, 180
        PSDI, 180
Limit switch, 136
Linear/rotary motion detection, 136

Machinery:
    benefits of systematic approach, 22
        design, 23
        maintenance/modification, 23
        specification, 22
    definition, 2–3
    distinction with process safety control
        systems, 7–8

engineering tasks, 19–21
example, 14
  development of integrated safety systems, 19
  equipment choices, 17
  programmable systems, 18–19
  proposed safety functions, 15–16
  risk assessment, 15–16
  risk reduction, 17
  standard solutions to standard problems, 18
hazard, 10
international standards/practices, 8–9
  owner's responsibility for safety, 10
  safety engineering methods in process plants, 8
  supplier's responsibility for safety, 9
risk, 10
  reduction, 11–12, 17
safety system, 3
  essential elements, 5
  false commands, 4
  fixed guards, 4
  passive guards, 4
  simple control scheme, 6–7
  sources of command, 4
Machinery concept hazard analysis (MCHA), 67
  activity analysis, 68
  analysis process, 67–8
  list of hazards, 67
  preparing for analysis, 67
Maintenance work, 21, 23
Mechanical limit switches, 144
  cam-operated, 145
    advantages, 145
    disadvantages, 145
    features, 145
  hinge-actuated, 145
    advantages, 145
    disadvantages, 145
  tongue-operated:
    advantages, 144
    disadvantages, 144
    features, 144
Mechanical pull out devices, 175
Mechanical trapped key interlocking, 149,
  150, 152
  application example, 153–4
  combination systems, 160
  key exchange boxes, 152
  limitations, 154
  principle behind, 152
  products, 152
  rotation sensor units, 152
  safety key, 152
  search and seal system, 153–4

  solenoid-controlled, 152
  time delay units, 152–3
Mechanical trip switches, 154–5
Monitoring relays, 125–7
  devices:
    IFM Electronic range of safety-related control
      systems, 129–30
    Moeller Electric range of protective
      components, 128–9
  functions, 128
  using relays with PLCs, 127
  using safety relays for guards, 128
Motion detection, 151
  safety design reminder, 151

Non-contact (non-mechanical) actuation, 146
  magnetically actuated switches, 146
  reed switches, 146
  Sipha 6 control unit, 146
  use of catalogs, 146

Opto-electronic device, 132, 158–9

Point of operation devices, 173
  remote controls, 173
  two-hand controls, 173–5
    benefits, 174
    disadvantages, 174–5
    OHSA standard, 174
Point of presence detectors, 175
  OHSA standard, 175–6
Presence sensing devices, 154
  AOPD guarding, 162
    access, 162
    area/perimeter guarding, 162
    point of operation, 162
  edge detectors, 157–8
  laser scan sensors, 163
  light curtains, 160
    safety features, 160–1
  mechanical trip switches, 155
    trip wires, 155
  opto-electronic detectors, 158
    general principles/types for safety, 158–9
    operating principles, 158–9
  pressure-sensitive mats, 156
    CKP/Solo safety mat, 156–7
    features, 156
  trip, 134, 154–5
    importance of 'safety distance', 155
Pressure-sensitive safety mats, 156, 176
  benefits, 176
  CKP/Solo, 156–7

Pressure-sensitive safety mats *(Continued)*
    disadvantages, 176
    features, 156
Pressure-sensitive strips, 136
Programmable E-stop monitors, 127–8
Programmable systems (PLCs), 190
    application software, 214–15
    automation safety, 18–19
    benefits, 195–6
        cost-effectiveness, 196
        performance, 195
        productivity improvements, 196
    characteristics, 201
        1oo2PLC example: Siemens 95F, 206–7
        1oo2PLC other examples, 207
        common cause potentials in redundant
            PLCs, 209
        design, 202
        diagnostic coverage, 201
        dual-channel, 204–5
        fault-tolerant time, 203–4
        hardware, 201
        larger 1oo2 systems, 207–9
        PILZ PSS key features, 211–13
        PILZ PSS safety PLC: 1oo3 architecture, 210
        series-connected, 205
        single-channel architecture with diagnostics,
            202–3
        single-channel safety, 207
        software, 202
        summary, 213–14
        triple modular redundant (TMR) systems,
            209–10
    classification/certification, 218–19
    design position, 93
    development
        attractions,196
        problems, 197
    disadvantages, 196–7
    I/O stage diagnostics, 197–9
        overall reliability, 199
    interfacing to the safety PLC, 193
    option of separate PLCs for control/safety, 193
    options for machinery control system, 193
    principles/use in safety control functions, 191–2
    safe networking, 215
        AS interface bus, 216–17
        general practices for safety networks, 218
        PILZ safetyBUS, 215–16
        profisafe/Profibus, 218
    software reliability considerations, 200
    summary, 219
    upgrading, 200–1
        using integrated control/safety system, 194
        using safety PLC for all controls, 194
Programmable systems (standards), 220
    outline of IEC 61508, 221
        competence of persons, 225
        contents, 222
        documentation, 225
        functional safety, 221–2
        key elements, 225
        management/design of functional safety, 225
        problem with safety categories, 225
        relevance to machinery safety, 225
        safety lifecycle, 223–4
        technical requirements, 225
    SILS:
        concept, 226
        determining required SIL for safety function,
            228–9
        implications for machinery systems, 230–2
        performance/design features definition, 227
        ratings comparison with safety categories,
            227–8
        role of IEC 62061, 230–2
    summary, 232
    see also regulations/standards
Protection, 169–70
    choosing, 170
        E-stop devices, 171
        physical guarding, 170
        safeguarding devices, 171
        selection factors, 171
    guarding devices:
        movable/automated, 172–3
        point of operation, 173
        selection factors, 171
    methods, 135–6
    point of operation devices, 173
        edge-sensitive devices, 177
        light curtains, 177–89
        mechanical pull out devices, 175
        point of presence detectors, 175–6
        pressure-sensitive safety mats, 176
        remote controls, 173
        two-hand controls, 173–5
    selection, 135

Regulations/standards, 24
    conformity procedures, 39
        Annex IV listed machines, 41
        contents of technical file, 40–1
        declaration of conformity, 39–40
        declaration of incorporation, 40
        EC type examination, 42

non-listed machines, 39
safety components in Annex IV, 42
UK implementation, 42–3
UK safety regulations for supply of
machinery, 43
EMC Directive, 45
limitations, 46
vs machinery Directive, 46
history/overview, 25–6
CE mark, 28, 30
harmonized, 31–3
knowing which apply, 29–30
principles, 27–8
relevance to workshop, 27
relevant to machinery safety, 26–7
supply side laws/New Approach, 28–9
technical requirements, 31
Type A, 32
Type B, 32
Type C, 343
international, 8–10
Low Voltage Directive, 44
Machinery Directive, 33
Annex I/EHSRs, 35–6
Annex IV/special categories, 38–9
application, 34
contents/structure, 35
finding, 34
linking Annex I EHSRs to standards, 36–8
relationship between regulations/standards, 38
role of type C product standards, 38
MD or LVD, 45
practices/enforcement by law, 25
safety networks/sensors, 166
supply side summary, 46
Type A – basic, 32, 50
Type B – group, 32, 50–1
Type C – product, 33, 51
USA, 52
application standards, 53
control reliability/foundation of safety
systems, 54–5
design standards, 54
incorporation of safety standards into law, 52–3
key OHSA regulations, 53
regulatory bodies, 52
user side directives:
PUWER, 48–9
responsibilities of user as buyer, 48–9
UK implementation of Framework Directive,
47–8
workplace health and safety, 46–7
see also programmable systems (standards)

Risk:
definition, 10
elements, 72
ability to maintain safety measures, 73
exposure/effects relationship, 73
human factors, 73
information for use, 73–4
persons exposed, 73
possibility to defeat/circumvent safety
measures, 73
reliability of safety functions, 73
type, duration, frequent of
exposure, 73
management, 1
measurement, 74–5
catastrophic, 74
fatal consequences for one person, 74
frequency/likelihood, 74
major, 74
minor, 74
quantitative, 74
tolerable, 12–13, 78
ALARP principle, 12–13
unacceptable, 11
Risk assessment, 13–15, 16, 56–7, 90
after adding protection measures, 16
determine limits of machine, 59–60
documentation methods, 90–1
checklist, 90
Dyadem, 91
Safexpert, 91
software tools, 91
example of intended/foreseeable misuse of
machine, 60–2
outcomes:
achieved objectives, 89
comparison of risk, 90
final evaluation, 89
practical example, 85–7
re-assess safety of machine, 87–9
procedure, 13, 57, 58–9
who should do it, 57
Risk estimation, 71
Guardmaster method, 77–8
risk bands, 78
matrix, 75
measurement, 74–5
PILZ method, 76–7
published risk scales for machinery
practice, 76
ranking scales, 75
tolerable risk concept, 78
using scales, 78

Risk reduction:
   application of layers protection, 83–4
   design improvements, 11
   evaluation of expected, 17
      consider failure modes/limitations of
         protection measures, 17
   principles, 79
      by additional precautions, 83
      by design, 79–81
      by information for use, 82
      by safeguarding, 81–2
   unacceptable risk, 11

Safeguards, 171
   guards/safety devices distinction, 134–5
   see also guards/barriers
Safety:
   categories, 104–5
      block diagram models, 106–8
      guidance for selecting, 108–10
      summary, 104–5
      table, 105–6
   control functions, 101
      e-stop function, 102
      fluctuations, loss, restoration of power
         resources, 104
      local control function, 103
      manual reset, 102
      manual suspension of safety functions, 103
      muting, 103
      response time, 102
      safety-related parameters, 102–3
      start/restart, 102
      stop function initiated by protective
         device, 101
   electrical, 1
   functional, 1
   guard sensor issues, 146
      by design, 147
      by redundancy, 147–8
   mechanical, 1
   physical, 1
Safety gate monitors, 128
Safety-instrumented system (SIS), 7–8
Safety integrated systems, 19
Safety integrity levels (SILs), 225
   concept, 226
   determination for machinery safety application,
      234–9
   performance/design features, 227
   ratings comparison with safety categories,
      227–8
   requirements, 228–9

Safety key switch, 136
Safety networks/sensors, 166–8
   standards, 167
Safety-related electrical control systems
      (SRECS), 1, 20
Safety relays:
   E-stop, 116–17
   monitoring, 120–4
      using with PLCs, 127
   practical, 117–18
      adding additional interlocks, 118–19
      category 3 E-stop monitoring, 120–2
      category 4 E-stop Banner Engineering
         example, 123–4
      category 4 E-stop monitoring, 122–3
      expander modules, 124–5
      monitoring of final elements, 120
      monitoring of valves, 120
      requirements for the reset switch, 124
      warning note, 124
   product description, 115–16
   terminology, 114
      positively guided contacts, 114–15
Sensors, 132–3
   control devices for safety, 163
      E-stop devices, 165–6
      hazards of pushbuttons, 164
      two-hand controls, 163–4
   devices for guards, 144
      guard-locking interlock + limit switches,
         149–50
      guard-locking systems/devices, 148–9
      mechanical limit switches, 144–5
      non-contact (non-mechanical)
         actuation, 146
      safety issues for guard locks, 150–1
      sensors for motion detection, 151
      suggested references, 146
      summary of safety issues, 146–8
   mechanical trapped key interlocking, 152–3
      application example, 153–4
      limitations, 154
   presence sensing devices, 154
      AOPD guarding, 162–3
      edge detectors, 157–8
      importance of 'safety-distance', 155
      laser scan for area guarding, 163
      light curtains, 160–1
      mechanical trip switches , 155
      opto-electronic detectors, 158–9
      pressure-sensitive mats, 156–7
      trip devices, 154
   protection of people, 139

review of guards, 138
  adjustable, 140
  automatic push-away guards, 140
  design requirements, 138
  distance, 140
  electrical control interlocking, 143
  features of interlocking, 142–3
  fixed, 138–40
  movable/interlocking, 140–2
  safe by position, 140
safeguarding types, 133–4
  control devices/sensing trip devices
      distinction, 134
  guards, 134
  identification, 135–7
  safety devices, 134
  selection of protection method, 135
  sensors/protection types relationship, 134–5
safety networks/sensors, 166–8
  advantages, 168
  E-stop controversy, 168
Solenoid lock, 136
Standards *see* regulations/standards
Stop functions:
  certification, 125

definitions/implications, 112–16
electronic/programmable E-stop monitors, 127–8
E-stop, 102, 112–13, 165–6, 171
  controversy, 168
  switching off, 112
initiated by protective device, 101
monitoring relays, 124–6
  other functions, 127
  product guides, 130
  safety guards, 128
  using with PLCs, 127
safety relay terminology, 114
  E-stop in practice, 116–17
  positively guided contacts, 114–15
  practical, 117–18
  typical product description, 115–16
summary, 132
Suppliers:
  responsibilities, 43
  UK safety regulations, 43

Trapped key switches, 136
Trip devices, 134, 154
Triple modular redundant (TMR) systems, 209–10
Two-hand control device, 136, 163, 173–5